KT-166-656

We Now Disrupt This Broadcast

We Now Disrupt This Broadcast

How Cable Transformed Television and the Internet Revolutionized It All

Amanda D. Lotz

The MIT Press
Cambridge, Massachusetts
London, England

This book was set in ITC Stone Sans Std and ITC Stone Serif Std by Toppan Best-set Premedia Limited. Printed and bound in the United States of America.

Names: Lotz, Amanda D., author. | Landgraf, John, writer of foreword.
Title: We now disrupt this broadcast : how cable transformed television and the internet revolutionized it all / Amanda D. Lotz.
Description: Cambridge, MA : The MIT Press, [2018] | Includes bibliographical references and index.
Identifiers: LCCN 2017032611 | ISBN 9780262037679 (hardcover : alk. paper)
Subjects: LCSH: Television broadcasting–United States. | Cable television–United States. | Television–Technological innovations. | Television broadcasting–Technological innovations. | Internet television–United States.
Classification: LCC PN1992.3.U5 L685 2018 | DDC 384.55/32–dc23 LC record available at https://lccn.loc.gov/2017032611

10 9 8 7 6 5 4 3 2 1

For Calla and Sayre
Because, why not?

Contents

Foreword

Once in a while, conditions arise in which artists find themselves with the institutional support to be truly brave, to resolutely follow their muses in the pursuit of truth and beauty. These tend to be eras in which new ideas and insurgent institutions are rising—times when we are emboldened to follow the bravest explorers as our tastes change. As economic competitors rush to meet a new demand, artists may find themselves, for a while, with the financial support to hold fast to their highest aspirations. We call these hallowed periods "Golden Ages." In this book, Amanda Lotz has written a history of the forces and conditions that have given rise the recent Golden Age of Television.

It was the advent of new channels in television whose subscription-driven business models could be powered by "great television" (instead of "most popular television") that ushered in this age. Originally, the movement was sparked by HBO and their flagship series *Sex in the City* (1998), *The Sopranos* (1999), and *The Wire* (2002). FX was the first basic cable channel to join the revolution, with the launch of *The Shield* (2003), *Nip/Tuck* (2004), and *Rescue Me* (2005). Showtime launched *Dexter* (2006), and then AMC launched *Mad Men* (2007) and *Breaking Bad* (2008). By this time, the floodgates opened and many more channels, and later streaming services, joined the party.

Of course great television existed before this Golden Age, just as a few works of art can find their way through the cracks in all times and places. Grant Tinker coined the phrase "first be best, then be first" when he was running NBC, a broadcast network that at the time was one of only three U.S. channels airing a high volume of scripted television series. And there are plenty of earlier shows, from *Playhouse 90* and *The Twilight Zone* to *All in the Family* and *Hill Street Blues*, which were the progenitors of more recent great series like *The Sopranos* and *Louie*.

But broadcast television's first priority was ratings, not acclaim. This was an inevitable outgrowth of the fact that its profit was driven by advertising, which in turn was driven by ratings. This priority—that overall popularity trumps excellence—meant that great television, even though it existed, was the exception. Over the last twenty years, as businesses have tried to lure new subscribing customers amid a growing abundance of choice, it became the rule.

In 2017, the TV business is bigger and television programming is better than it has ever been. As we come to the close of this year, there will be about 500 original scripted television series produced for the U.S market, more programs than at any time in the history of the medium—and a number that has more than tripled in the past two decades. A surprising percentage of these 500 series will be excellent, artistically ambitious shows. I've been fortunate enough to attend an annual American Film Institute luncheon for many years, and event in which the ten best American films and ten best American television series are celebrated. This year, FX will receive its twenty-second AFI top ten season recognition. When I first started attending, the films were (on average) better than the TV series. Sometimes one felt the jury had to scrape the barrel a little to come up with ten TV series that they could proclaim to be great. But something changed. Over the past decade, the barrel scraping was actually more often needed to come up with ten great films than ten great TV series. When Newton Minnow, the former chairman of the FCC, referred to TV as "a vast wasteland" in 1961, I doubt any sane person would have believed such a situation was even theoretically possible. But, as Professor Lotz describes in this book, a set of new business circumstances, made possible by a change in television distribution, led to something remarkable.

At the semiannual Television Critics Association summer press tour three years ago, I used the phrase "Peak TV" to try and provide some context to the unprecedented amount of scripted programming being produced. At that point, I was already starting to get an uneasy feeling about where our Golden Age might be heading. I naively predicted at that time that the amount of scripted shows would reach an apex in 2017. The following year, I had to revise that prediction, because I underestimated the arms race that would be fueled by the entry of OTT players like Netflix, Amazon, and Hulu (and soon Apple and Facebook), with their multibillion-dollar war chests.

By now, I'd be hard pressed to forecast when the present orgy of spending on television might end, so instead I will make a different prediction: that from a creative standpoint, more will not be more. Despite some excellent programming contributions made by these new streaming services, what we are witnessing is the transition from our precious Golden Age into an overheated economic bubble. TV series, once again led by HBO with their excellent, epic *Game of Thrones*, have now become the primary weapons in a war for global domination for subscribers and internet users. And, unfortunately, though weapons can sometimes be quite beautiful, they tend to be something other than art. Those whose first goal is scale or profit do not rate the artist's search for truth and beauty as their highest priority (even though they might admire it).

And so we find ourselves at a crossroads. Will those that control the financing of television, driven by their titanic struggle over who owns the future, kill the Golden Goose to more quickly extract the even greater riches we believe must be inside? Or will the artists be able to keep their feet on the ground and their heads in the clouds for a while longer?

In his book *Sapiens: A Brief History of Human Kind*, author Yuval Noah Harari posits the theory that after three million years of using fire, making tools, wearing clothes, living and hunting in bands, and spreading out across the entire contiguous land mass of Africa, Europe, the Middle East, Russia, and Asia, our ancestors made a huge leap forward when they mysteriously gained an ability that had never before existed on our planet: the ability to imagine, to create things in our mind that do not exist. The author suggests that this mysterious ability gave rise to human civilization. Cities and religions and empires rose for the first time on the foundation of stories that could inspire and unite thousands or millions or even tens of millions of people.

As Harari says, chimps and dolphins and elephants live socially and coordinate well in bands numbering in the dozens. But if you could theoretically get a million chimps together in the same place, one thing it would surely not look like is the Olympic Games. Only stories that compel us (as the Olympics do) can unify and coordinate humans in the millions. Harari's book theorizes that every key technological innovation, from the domestication of plants and animals to the industrial revolution, came about as a result of larger populations of humans beginning to live together because we had stories and foundational myths to unite us—not the other way around. Think about the profound implications of that basic equation:

technology—farming, industry, books, computers, the internet—does not beget the stories that follow upon its development. It is the stories that beget technology.

So humans are, quite literally, the ANIMAL WHO TELLS STORIES. And imagination is the foundational tool of our cultural identity as well as our dominance on this planet.

If stories beget technology, and not the other way around, we face an interesting crisis as companies driven by technology now begin to conquer and control the artists who are their own progenitors. We all stand in awe of what technology has done to make our lives easier. But just as the hero must endure pain, overcome his own inner demons, and vanquish his foes in any great saga, each of us needs discomfort, inconvenience, and opposition to reach our full potential. Even as I type this using a wonderful word processing program on a sleek laptop, I fear the frictionless world that technology is creating will impoverish us and our stories us in ways we do not yet understand.

I've been extremely fortunate over the past fifteen years to run a television brand during a period when I could best serve the needs of the business and its shareholders by first serving the needs of the artists who chose to make their home at FX. I have an undergraduate degree in anthropology, not business, and as a person who has spent his life celebrating the richness of human stories, I enter the next television age sincerely hoping that whoever decides what gets made won't subjugate our stories—the source of our humanity—to giant, data-driven machines and the ambitions of the vast new corporate empires they enable.

When we look back twenty years from now, perhaps Professor Lotz's incisive work will describe the creation of a Golden Age that came to a close, and we will look back at it with forlorn nostalgia. Or maybe I am wrong, and we will instead find that the past twenty years were just a prelude, creating the foundation from which even greater creative heights would be scaled.

Isn't it the ultimate privilege, and the definition of life itself, to live inside a story whose end we do not know?

John Landgraf
CEO, FX Networks and FX Productions
Los Angeles, CA

Preface

The medium we call television has undergone many changes in recent years, but it is important to remember that change has been a constant during the relatively brief time—only a bit more than a half-century—that it has been in existence. My earliest memories include the advent of cable—the last great disruption in the development of television. Most particularly, I remember a toggle switch attached to our set that I had to slide from position A to position B in order to watch what I simply knew then as "more channels." Only in college media classes did I learn the different rules and business models that distinguish broadcast networks and cable channels. Accordingly, for two decades now I've tried to explain the difference between these two sectors of the television industry to students who never even had the experience of sliding the toggle from A to B to watch MTV instead of NBC.

A lot has changed since those days in the early 1980s, but those norms were still largely in place in 1996, when *We Now Disrupt This Broadcast* begins its story. Television shows have changed a lot during the fifteen years I've been writing about the television industry, and so too have conversations about television. Television was still held in disdain by the arbiters of popular culture when the story this book tells begins. Watching television was not something "educated" people readily admitted—they'd be far more likely to suggest their sophistication by touting that they didn't even own a television.

The changes in the business of television explained here expanded its creative—yet still commercial—possibilities. The adjustments are wide-ranging. Top Hollywood creative talent that would never have considered working in television twenty years ago can be found on screens any night of the week. Cultural opinion leaders such as the *New York Times* can't publish enough about the shows of the era, and the *Times* even has made television the center of one of its first newsletters/recommendation sites, "Watching."

The television shows of the last two decades have been so important that many journalists have written about them continually in newspapers, magazines, and a growing array of web publications as well as in several books that take television seriously and focus on its accomplishments—particularly Alan Sepinwall's *The Revolution Was Televised*, but also Sepinwall and Matt Zoller Seitz's *TV: The Book*, David Bianculli's *The Platinum Age of Television*, and Brett Martin's *Difficult Men*. *We Now Disrupt This Broadcast* is a companion to these books, which celebrate the television of the last twenty years and discuss its series in great depth. Rather than rehash the thoughtful examination of particular series they provide, this book explains the behind-the-scenes shifts in the business of television that enabled these shows to be created.

The story of U.S. television's transformation should prove fascinating for readers with many different interests. Those who have been fans of shows such as *The Sopranos*, *The Shield*, and *The Walking Dead* can learn about the transformations in the business of television that made such once-unimaginable series possible. Those curious about headlines claiming the impending "death" of television and who find that Netflix and Amazon Video now provide much of their television viewing can learn about how different the business of television is now than it was in even the recent past. Other readers may be interested in changing media businesses and eager to read a case study of how television has survived and even thrived despite the massive disruption of competitive norms and the emergence of strong new competitors.

This book, with its story of how and why the business of television changed so profoundly between 1996 and 2016, has much to offer all of those readers. The main text tells the general story of this change. The book also includes extensive notes that provide more detail and nuance and dig into the deeper complexity of the industry and its practices. These notes are intended for people who work in media industries or those studying various aspects of media, and they offer a closer look into intricacies of business practice.

The research supporting *We Now Disrupt This Broadcast* is also the basis of several publications targeted to academic audiences. For readers seeking even greater nuance, please see:

Lotz, Amanda D. "The Paradigmatic Evolution of U.S. Television and the Emergence of Internet-Distributed Television" *Icono (Journal of Communication and Emergent Technologies*) 14, no. 2 (2016): 122–142. doi: 10.7195/ri14.v14i1.993

Lotz, Amanda D. *Portals: A Treatise on Internet-Distributed Television.* Ann Arbor: Maize Publishing, 2017.

Lotz, Amanda D. "Linking Industrial and Creative Change in 21st-Century U.S. Television." *Media International Australia* 164, no. 1 (May 11, 2017): 10–20, special issue: "TV Now," edited by Sue Turnbull, Marion McCutcheon, and Amanda D. Lotz.

Acknowledgments

This book is a long answer to a question Russ Neuman posed over lunch in early 2006. I was working out ideas for my book *The Television Will Be Revolutionized* (2007) and Russ insisted that cable television wasn't important, just an expanse of "rerun" broadcast shows. My answer at the time was a frustrated, "yes, but." Given my occupation, I had been watching every new scripted series released by a cable channel and knew the norms of the industry were starting to change, but these shifts weren't more generally obvious. It has taken a long time to form a fuller reply, but uncovering the mystery of how and why cable changed has been a rewarding adventure.

The second half of this book grew out of questions raised during my participation in the Connected Viewing Initiative of the Media Industries Project in 2012. Without that opportunity I might not have pivoted into questions about the implications of internet distribution for television. My thanks to Michael Curtin and Jennifer Holt for providing that opportunity and to Ethan Tussey for thoughtful encouragement of half-formed ideas.

I had many helpful conversations along the way of figuring out the story of this book and conventional interviews with others once the story started taking shape. I'm deeply grateful to the following for generously sharing their time and insight: Andy Breckman, Kevin Beggs, Peter Benedik, Jackie de Crinis, Tom Fontana, Marc Graboff, Cathy Hetzel, Jordan Levin, Jon Mandel, Nancy Miller, Rod Perth, Daniel Pipski, Shawn Ryan, Robin Schwartz, Evan Shapiro, and Anne Thomopolous. My thanks as well to Jim Burnstein, Scott Weiner, and Jordan Levin who helped me make many of these connections.

My thanks to Jonathan Gray for reading an early draft and talking me down more than a few times. Trying to transition into trade writing (and publishing) was a lot like hitting my head against the wall for months on end. I appreciate the counseling offered by Anne Trubek and the University of Michigan ADVANCE Summer Writing Grant that allowed me to hire her as a writing coach. Kitior Ngu and Annemarie Navar-Gill provided excellent research assistance only possible because of a grant from the University of Michigan Associate Professor Support Fund. Horace Newcomb proved mentorship never ends and generously provided late stage suggestions. Thanks to Jesús Hernández for an unforgettable title, Gita Manaktala for support of some unconventional choices, Michael Sims for an excellent production experience, and also to Brian Kenny at the Cable Center for help retrieving needed material.

There is now a deep chorus of academic colleagues whose voices I hear when I write and edit. I have debated you at length (in my head). I've doubtlessly failed to assuage your criticisms completely, but you make my work stronger, and I appreciate that. Before you become angry with me, do read the notes. It isn't possible to do all the typical academic things when not writing a typical academic book. I ask for your indulgence; I needed to try to master a new skill.

I offer heartfelt thanks as well to the many friends and colleagues who helped me through various stages of frustration that came from insisting on trying to write this book for a broader audience. This task fell mostly to Wes Huffstutter, who tolerated a lot of manuscript-derived crankiness. Thank you, as always, for your support and patience.

Part One Cable Transforms Television, 1996–2010

1 Transformation, Then Revolution

Pick your favorite television show from the last twenty years. Odds are that changes that occurred in the business of television over the last two decades were responsible for its existence. Or imagine waking up from a twenty-year coma and having the changes in television explained. NBC's legendary "Must-See-TV" Thursday night line-up—*Friends*, *Seinfeld*, and *ER*—has been replaced by a new breed of program: *The Sopranos*, *Mad Men*, and *Game of Thrones*, not to mention new service providers such as Netflix. More than just changes in shows, television watching behaviors unimaginable twenty years ago are now common: you can watch a show when you want, rather than when a network schedules it; you can download episodes onto a tablet or watch them live on a mobile phone wherever you receive a signal.

Anyone living in the United States over the last twenty years knows that television has changed tremendously. Just how great a change you perceive depends largely on how old you are. But whether you are eighteen or eighty-one, there is no denying that prime-time television in 2016 bore little resemblance to the television of just twenty years earlier.

We didn't know it at the time, but 1996 marked the end of an era. These were the last days in which it was normal for television shows to gather a large and diverse mass audience. By 1996, new broadcast networks such as The WB and UPN—which merged to become The CW in 2006—were siphoning viewers from "the Big Three" networks: ABC, CBS, and NBC. These organizations had spent forty years as the only broadcast networks. Fox launched in 1986 and its fast growth led to accounts of the "Big Four" within a decade. Cable channels attracted increasing numbers of viewers as well. By the last years of the century, the first digital video recorders (DVRs) entered American homes and the pace of technological change grew even faster: digital cable, video on demand, and internet streaming.

In the blink of an eye, television transitioned from a medium with programs for mass audiences to one that spoke to an array of niche tastes and interests. Of course, a few events still draw a mass audience to the same television channel at the same time, but it is the exception rather than the norm. What is important—and contrary to many accounts—is that the business of television did not weaken amid these transitions. But it certainly changed.

For most viewers, this change led to the development of shows they never imagined. For many, expectations of what U.S. television could be or do was altered forever in 1999 with the launch of *The Sopranos*. A quick review of some headlines makes clear its legendary status: "*The Sopranos* Is the Most Influential Television Drama Ever"; "The Best Written TV Show: *The Sopranos*, Of Course"; "Revolution on the Small Screen: How Dramas Like *The Sopranos* Transformed TV."[1] Even those who never watched *The Sopranos* knew about it.

Despite these claims, this book tells the story of television's transformation without attending much at all to *The Sopranos*. *The Sopranos's* exceptional success has overshadowed consideration of the role of series that prepared its possibility. Rather than retelling stories about this well-known series, the book begins with those just before it.

Many recognize U.S. television's considerable change, but few know why it shifted so significantly. Behind the screens that brought us *Sex and the City*, *Breaking Bad*, *The Walking Dead*, and *Stranger Things*, the business of television also was transformed. Television evolved from competition among just a few dozen channels to a few hundred. Even more revolutionary change then developed in response to competitors such as Netflix, Hulu, and Amazon Video that distributed television over the internet. What viewers saw on screen may have been jaw dropping, but behind the scenes, the transformation for television executives was often gut wrenching.

The early twenty-first century was full of pronouncements about the death of television and the rise of "new media," but the surprising protagonist in this story of change is "cable."

Cable? How could a twentieth-century technology be the lead character in television's revolution? The story of how and why U.S. television changed so profoundly at the beginning of the twenty-first century is a history in two parts. The first explores how cable channels moved from ancillary, second-tier status to television's innovative core and profit center.

The second begins in 2010 as internet distribution of video disrupted both cable and broadcast channels.

At the beginning of the 1990s, cable programming was bad enough that it was the butt of jokes. Jackie de Crinis, a senior vice president at the USA cable channel in the early 2000s, recalls, "No one would come to us. We couldn't afford showrunners whose names anyone knew."

The considerable change that has taken place over just two decades has led American audiences to forget the recent past of cable channels as a backwater of "rerun" broadcast network series and old movies. After decades of secondary status, cable channels rose abruptly to prominence and became regarded on a par with broadcast television. This produced wide-ranging consequences throughout the media landscape. By the end of the decade, cable channels had created innovative shows such as *The Shield*, *Sons of Anarchy*, and *The Closer*. Cable dramas dominated television awards as well as conversations about television, and they were rewriting expectations of television around the globe.

Many initially imagined "the internet" as a new medium, akin to television, film, or magazines. The industry, tech, and even mainstream press opined endlessly about the coming death of television throughout the early twenty-first century, predicting its assassination by the internet.

But the biggest impact of the internet for television was as a new way to transmit programming. Rather than killing television, internet distribution markedly improved how we watch. Crucially, internet distribution has capabilities that broadcasting, cable, and satellite lack. Internet-distributed television allows audiences on-demand access. This capability is forcing a reinvention of television businesses that were built on scheduling programs. Changes in the business of television change what shows are made and what audiences can watch.[2]

The story of how evolution gave way to revolution is a story not of death, but of a society's response to the collision of new technologies, changing business strategies, and unprecedented storytelling. Many obituaries were written for television in the first years of the twenty-first century, but a decade later—with television remaining very much alive—forecasters instead anticipated the death of cable.

These changes occurred so quickly that little understanding of these sizable transitions exists. How is it that cable—an industry that spent thirty years as the dark horse of U.S. television—became the source of programs that reinvented television as a creative form? How did cable service providers, such as Comcast, transform themselves into the internet providers for the majority of American homes?[3] Why didn't these changes transpire earlier, and what allowed them to take place when they did?

Scripted entertainment—television dramas and comedies—has been the programming most disrupted by new distribution technologies. Other types of television programming, such as news and sports, have not evolved as quickly, but they also face substantial disruption.

We commonly consider "cable" as both the programs of cable channels and the infrastructure of cable service providers, but they are notably different businesses. The technologies that deliver television to our homes and devices are a crucial part of this story. Distribution of signals over wire was always the core business of the cable industry, and the implications of the evolution of "cable" providers into internet providers is an overlooked story of the last two decades. Companies that began as cable operators, such as Comcast, Charter, and Cox, are often monopoly providers of the home internet service that has become a crucial resource of modern life. They are the "barons of the new gilded age," able to leverage vast resources to ensure their market power remains unchecked.[4]

Such attention to cable may seem surprising given the spotlight on internet-distributed services such as Netflix in recent years. As new organizations of video content emerge, cable channels begin to fade from importance—perhaps even out of existence. Yet the underlying system of wires that connects most homes to video communication and entertainment builds upon the skeleton of the cable industry. In their capacity as internet providers, cable companies now deliver Netflix, YouTube, and an array of internet-distributed television. Even if the future of cable is very different from its past, its story is most certainly not one of death. To begin to make sense of these developments, we must first understand that cable involves two separate and very different businesses.

2 Cable?

The word *cable* is used a lot like *Hollywood* in American culture. While it references something quite specific, it also stands in for much more and describes two separate entities. When we'd say, "I'm going to see what is on cable," we meant the range of *channels* delivered by the cable service. Cable, however, brought cable channels (ESPN, CNN, and MTV) into our homes as well as broadcast networks (ABC, CBS, NBC, Fox) that we could also receive for free with an antenna.

Even though most think of cable channels and certain programs when the word cable is mentioned, this has never been the core of the *cable industry*. The cable business has been foremost about distribution, or the business of the cable service providers. This is the cable we mean when we say, "I have to pay the cable bill." Here we are talking about the business of companies such as Comcast, Charter, and Cox, which have millions of customers around the country, but mostly in or near cities. There are also hundreds of small cable companies with just a few thousand subscribers that service smaller and more rural places.

For the most part, the companies we pay cable bills to are unrelated to those that provide the programs we enjoy. With the exception of Comcast, cable service providers own few channels and little content.[1] More often than not, the cable service providers find themselves at cross-purposes with cable channels. Both have profit models that rely on receiving more money from viewers' monthly payments. When viewers are unwilling to cough up more, an acrimonious battle ensues between these two cable sectors.

To give a sense of the differences between the cable channel and cable service businesses, in 2004, cable networks produced $34 billion of revenue, while cable service providers accounted for more than $57 billion.[2] What is important here is that cable service was a bigger industry—by $23 billion—even before internet service became a crucial component of services offered by the cable industry. The revenue produced by providing cable service—derived from the fees paid monthly by subscribers—was nearly *twice* that of cable channels, which earn revenue from selling advertising and from fees a cable service provider pays to include a channel in its service. In 2016, cable channels reported $75 billion of revenue and $15.7 billion in profit, while cable providers amassed $108 billion in revenue and $21.5 billion in profits.[3] In 2014, the profits of U.S. cable providers were greater than those of the country's major music labels, broadcast networks, newspapers, magazines, film production, and cable networks combined.[4]

Just as our contemporary understanding of cable often conflates cable channels and cable service providers, the history of cable in the United States is a complicated, contradictory, and confused tale in itself. And the history of cable in the United States is uncommon. The wired television history of the United States has been replicated few places elsewhere—only Denmark, the Netherlands, and Argentina were wired by cable to a similar extent. In most places, broadcast distribution gave way to satellite distribution rather than cable. That has been unimportant until recently, but it becomes significant in the second part of the story because the wires used by cable service providers enable their transformation into internet service providers.

Cable first developed to solve a problem of geography—the inability of towns, typically in mountainous areas, to receive broadcast signals. Powerful broadcasters avoided competition for decades by lobbying the

Table 2.1
Cable channel versus cable service industry revenue

	Cable channel	Cable service
2000	$20.9B	$47.9B
2005	$45.0B	$65.9B
2010	$60.4B	$87.7B
2015	$74.1B	$105.5B

*Source IBIS Reports 51321, 51322, December 2004; 51521, March 2016; 51711a, May 2016.

government to establish rules that forbade cable service in major metropolitan areas. Rules also prevented cable services from transmitting anything other than local broadcast stations. Cable was consequently born, and came of age, in small, in-between places tucked among mountains, such as Astoria, Oregon, Tuckerman, Arkansas, and Meadville, Pennsylvania. The cable business was technically born in the late 1940s, but did nothing more than retransmit broadcast signals until the 1970s.[5]

The business of cable channels has evolved considerably. For quite a bit of cable history—until HBO's launch in 1972—cable was simply a service for retransmitting broadcast stations. There were no other channels or programming available with cable service. By the late 1980s, having cable was something like what having a gaming system is today: it was a technology that offered more options and was found in about half of U.S. homes. Cable service offered limited choice. Services offered only a few dozen channels and, until the late 1990s, most channels filled their schedules with reruns of broadcast shows and old movies. Still, cable afforded the option to watch an episode of the *Brady Bunch* instead of the nightly news offered by all three broadcast networks.

The number of homes with cable grew considerably through the 1980s—from 22 percent to 60 percent of homes with television—as cable was finally permitted in metropolitan areas.[6] Many familiar channels launched in the 1980s. But into the 1990s, it remained a secondary and subpar component of the television landscape.[7]

The fortunes of the cable industry have evolved considerably over the last sixty years. Cable spent much of its history as an underdog, kept at bay and prevented from providing significant competition through effective lobbying by broadcasters who curried favorable regulatory policies that preserved their dominant role in U.S. television. A video produced by the cable industry's trade association in 2006 characterized the industry's origins as that of a "rag-tag group of dreamers." This self-history was profoundly propagandistic, but the stories of early cable entrepreneurs are indeed amazing and provide antecedents for the nearly mythical tales of contemporary high-tech entrepreneurs that capture imaginations today like Facebook's Mark Zuckerberg, Google's Sergey Brin and Larry Page, or Netflix's Reed Hastings. The cable entrepreneurs weren't located in a common place like Silicon Valley, New York City, or Hollywood; they were spread throughout the country, because cable originated as a local and largely nonmetropolitan industry.

Although the cable industry still imagined itself as a "rag-tag group of dreamers" in the early 2000s, even by the 1990s this was undoubtedly no longer the case. It was clear by the mid-1990s that cable—the channels or the service—was not going away. Having cable had become a common part of American life. But behind the scenes, cable was in trouble.[8]

3 A Death Spiral?

Although it seemed typical at the time, 1996 proved to be a watershed year for U.S. television. Many important developments occurred that make it appropriate to use that year as our starting point. Changes in the television industry led to the reinvention of cable, and then all of television.

In an expansive November 1995 article titled "Cable TV: A Business in Transition," *Media Week* journalist Michael Burgi began by noting, "A relatively new industry will see more upheaval in 1996 than it has ever seen before."[1] Although hyperbole is common in such articles, from the vantage point of 2016, Burgi's claim actually appears understated.

The upheaval came primarily from competition, some new, some only anticipated. The cable service business was booming in the early 1990s. By 1996, roughly 65 percent of U.S. households paid for cable service to augment their video options beyond those of the "free TV" offered by broadcast networks. Cable providers gradually added new channels through the 1980s and 1990s, quickly filling the capacity of the analog cable systems. Cable providers operated as local monopolies and, without the moderating force of competition, increased the rates they charged subscribers many times that of inflation. Congress's General Accounting Office reported in 1989 that basic cable rates rose 29 percent over the previous two years.[2] Feeling the price gouge and lack of alternatives, citizens demanded government action.

Congress acted, passing legislation regulating rates in the early 1990s, but it did not want to continue the task of cable rate regulation. Using a broad piece of regulation known as the Telecommunications Act of 1996, Congress changed rules that had prevented competition between cable and telephone providers hoping to introduce market-regulating competition.

The act allowed cable providers to offer phone service and phone providers to offer video service. This competition would take nearly a decade to develop, but it was clear that the playing field of cable would no longer be the same.[3]

Separate from this attempt to create competition, by several measures 1996 can be considered the first year the cable industry did indeed face competition. DirecTV launched in 1994, although only two million households accessed programming using its new small satellites—called DBS, for direct broadcast satellite—in that year.[4] By 1996, a range of DBS competitors materialized, and the competition between cable and satellite began in earnest.[5]

This competition looked a bit different in 1996 than by 2010 or 2016. Satellite's use of a digital signal allowed greater channel capacity than the still-analog cable systems of the era. The digital signal also provided improved picture and sound quality.[6] In 1996, DBS provided a product technologically superior to cable. DirecTV promotional material touted 175 channels, while the packages cable subscribers chose averaged 47. Satellite's channel offerings weren't as robust as 175 might suggest: they included east and west coast feeds of broadcast networks and some cable channels, the first one hundred channels on DirecTV were pay-per-view movies, and about thirty of the channels were regional sports channels.

In terms of the economics of different services, satellite providers paid higher fees to the channels they carried than large cable operators did. This meant that satellite operators had higher programming costs. Nevertheless, satellite services often priced their service lower than cable to win customers, even though it diminished profit margins. It cost satellite services roughly $1.5 million to launch a satellite and secure its orbit position, but this infrastructure was significantly less costly than the wires and distribution plants that cable had to maintain.[7]

A budget-conscious consumer with the choice between cable and satellite faced a decision between slightly higher monthly cable fees and no technology costs versus satellite's lower monthly fees, larger channel array, and better picture quality, but substantial technology investment. In these early years, dishes cost $400, although some services offered them at half that price with a one- or two-year service commitment. Also, each television set needed its own dish, weather could interfere with the signal, and the dishes needed a direct line of sight to the south. The mix of costs and benefits led some cable subscribers to switch to satellite, although it was by no means a stampede.

The emerging satellite competition nudged cable operators toward system upgrades that were increasingly overdue. Implementing "digital cable" meant rebuilding cable's infrastructure to send digital instead of analog signals. This would allow systems to provide more channels and transmit higher-quality signals, and it would enable the two-way messaging necessary for on-demand service.

Reading press coverage of the cable industry from 1996 and 1997 produces a sense of déjà vu. The stories of promised technological innovation and coming interactivity are stunning, not because they underscore the repetition of innovations in a way that seems uncannily familiar nearly twenty years later, but because it is actually a discussion of the same innovation *still* being discussed two decades on. The forecast of cable's digital, interactive, on-demand future persisted for more than a quarter of a century

By 1996, digital cable had been a "coming" technology long enough that investors began doubting its arrival. As a result, 1996 was a do-or-die moment for the cable industry. DBS had begun competing, and it brought the long-mythic, three-hundred-channel universe with it. If cable was to compete, it needed a digital infrastructure. The required technology had been slow in developing and the challenge of introducing facilities upgrades was growing more formidable.

As industry journalist Price Coleman explained in December 1996, "Digital cable has encountered repeated launch delays, largely a result of problems with development and distribution of set-top box terminals. But the stumbling block for digital was as much economics as technology."[8] He outlined how, by several crucial measures, cable systems had waited too long to upgrade their facilities and had arrived at a situation that can be a death spiral for an industry: cable systems' spending on upgrades outpaced their cash flow, lack of cash required the companies to borrow to finance the upgrades, taking on higher debt (the borrowed money) led analysts to decrease the companies' credit ratings, and decreased credit ratings increased the cost of money, which made expensive upgrades that outpaced cash flow even more expensive. To further compound the crisis, the investment community had grown concerned about competition from satellite, it worried that telco[a] competition would be as profound as satellite, and it did not yet appreciate the potential of internet service as a new revenue stream for cable companies. Cable stocks plunged 10 to 40 percent in 1996.[9]

Although these upgrades would become the backbone of internet service, the focus of the upgrades remained on cable's video service and the expansion of cable systems' capacity. Based on the state of technology in the late 1990s, transitioning to digital was expected to add another 75 to 100 channels of capacity, but advancements in compression technology expanded capacity beyond these estimates.[10]

Cable providers needed added capacity for many reasons. Obviously, it allowed systems to carry more channels, but another provision of the Telecommunications Act of 1996 mandated that broadcasters transition to a more efficient digital signal to enable high-definition television. The limited channel capacity of analog cable would become even more dire if cable systems were to carry the larger high-definition channels. A single HD channel was estimated to require the same bandwidth as four to six standard definition signals (again, these estimates are based on late 1990s compression capabilities).[11]

At the end of 1997, the cable industry was still mired in a potentially fatal struggle to secure the capital to finance the upgrades. The industry had deployed only 10,000 to 20,000 digital set-top boxes, and each box cost operators $400. Significant supply chain problems emerged, and only one company manufactured the boxes, which further slowed the rollout.[12] The cable industry made an $80 billion investment in its digital infrastructure in the late 1990s.[13]

Through a combination of debt and subscriber fee hikes, the industry invested in the new technology and steadily introduced digital cable. Within five years, 19 million homes (or 26 percent of the 73.5 million cable subscribers) subscribed to digital cable services. DBS penetration grew to 17.6 million by that time, yielding nearly 37 million digital multichannel homes in 2002.[14]

If cable service providers hadn't built a digital infrastructure in the late 1990s, the second part of this book would likely tell a very different story.

a. *Telco* describes the companies that began as telephone providers and were now offering cable video and internet service. They compete with cable service providers and are part of the multichannel industry that also includes cable and satellite.

Satellite would have quickly surpassed cable without the upgrade, making the situation in the United States more like the norm in the rest of the world when being wired for internet service became important a decade later. The telephone companies would likely have challenged satellite as internet service grew in importance and would not have had much competition. Buoyed by internet revenue, the telephone companies would have been in a stronger position to offer video services that would compete with satellite.

It is unlikely, though, that the story would have been substantially divergent for consumers given the equivalently uncompetitive nature of wired telephony. This situation might have inspired greater regulatory action or earlier government involvement in establishing internet service as a crucial utility, which could have allowed either a more competitive or subsidized internet infrastructure.

Cable stocks were still down at the end of 2002 and they would soon fall farther. Skepticism about a long-heralded digital revolution weighed heavily on Wall Street analysts' assessments of cable. Although nearly 30 percent of cable subscribers had transitioned to digital cable by 2002, satellite had taken many subscribers and fears were high that telcos would take just as many when they began their services. An accounting scandal at Adelphia, the seventh largest cable provider, led to a close reassessment of the rest of the cable sector in 2003.

Wall Street continued to regard cable stocks poorly until the gains of the digital investment and the red herring of telco competition finally became evident around 2006. Cable service providers were punished by investors who thought they were on the verge of being killed off by "the internet." It wasn't yet clear that cable was positioned to be the dominant provider of home internet service, and that this would be a very lucrative business. Even the more savvy analysts who recognized cable's advantage in having put wires into more than half of American homes—and ready to connect to most others—underestimated the coming importance of internet service in their evaluations. From 2000 through 2005, Wall Street analysts predicted that only 40 percent of homes—*at most*—would ever purchase high-speed internet service.[15] It surpassed that level in 2006.

Although these various industry-changing forces—the Telecomm Act, competition from satellite, the upgrade to digital cable, the transition to high-definition—are easy to view separately at the distance provided by a decade or two, the experience in 1996 was more akin to being in the midst of an earthquake. Everything seemed on the verge of changing at once; no ground seemed certain, and it was difficult to see what changes were related and which were independent of others.[16]

By 1996, a nascent "new tech" industry also had begun to emerge in the greater San Francisco area. Some saw the profound change that was coming, but for most, it was difficult to blend the old with the new. Burgi's 1995 article on cable's coming transition concludes with a look at what was then state-of-the-art modem technology offered by @home, which was one of the first internet service providers; it was bankrupt by 2001. He asks the reader to imagine the coming future—just two years off: "Consider a typical @home system, circa 1998: You come home from work and turn on the tube, and among your choices are watching normal TV, ordering video-on-demand programming, checking out community info and e-mail on the local @home-branded online service, surfing the Web, even answering the phone."[17]

This was the beginning of modern digital life—the years when what are now taken-for-granted digital technologies started entering everyday life. But at this emergent moment, they could be understood only through the lens of the technologies that had come before. The first years of this technological revolution belonged to the armchair technologists who saw only adaptation of current technologies. This led to forecasts such as on-demand television and television screens that people would check email on (never mind that email barely existed). But the dreamers envisioned the longer arc of technologies. Even midway into the first decade of the twenty-first century, only a dreamer could imagine watching television on a screen carried in your pocket. Or that the screen would provide stories viewers would be eager to watch.

4 300 Channels, Why Is Nothing On?

By transitioning to digital cable, the cable industry survived a near-death experience, but most viewers hardly noticed it. Although digital cable boxes began finding their way into American homes in the late 1990s, rollout was slow. The introduction of digital cable was crucial to the cable industry's future; its greatest opportunity, however—on-demand video—was a decade away from including the programs viewers wanted to watch. By 2016, "interactive" television hadn't materialized, but the arrival of the web had reduced the need.

It is worth taking a moment to recall just how different cable channel programming was in the late 1990s. Before 1996, a cable channel's value proposition was primarily that it provided an alternative to broadcasters' consistent fare. Broadcasters competed with each other, but in the narrowest sense. Each offered the same types of programming throughout the day: talk, game shows, soap operas, news, and similar prime-time entertainment.

Broadcasters had changed little in response to the growing cable sector by 1996, and with good cause. Relative to programming norms of the early 2000s, 1990s' cable programming was quite limited, and cable did not yet compete with broadcasters by offering programs of comparable budget or aspiration. Despite the rapidly expanding range of channels, cable offered little that was particularly new or innovative. By the early 1990s, ESPN, CNN, and Nickelodeon had clearly established identities and regularly drew substantial audiences, although these were still fairly early days: ESPN had *SportsCenter* and secured a major-league baseball license in 1990, but it was not yet the platform for major sports competitions that it became by 2000; and CNN derived its primary value as an access point for video news at any hour of the day. The success of these content-specific, niche channels did little to threaten broadcast dominance.

Cable and broadcast were different in important ways—particularly in their revenue model. Most cable channels derived revenue from both advertising and subscriber fees while broadcasters had only advertising. This explains a lot of the difference in programming and programming strategy of each sector in these years. Journalists covered the two sectors as a horse race between cable and broadcast, but emphasizing them as competing sectors mistook the competitive dynamic.

Even though different revenue strategies distinguished broadcast and cable, the same companies that owned the broadcast networks owned many of the most successful cable channels. It soon became the case that ESPN could draw more viewers to its football games than those aired by NBC. But this can hardly be considered a victory of cable over broadcast; ESPN is owned by a company (Disney) that also owns a broadcast network among several other cable channels, and NBC is similarly a cog in a conglomerate with varied cable properties. Once broadcast networks and cable channels became commonly owned—a process that transpired throughout the mid-1990s—seeing cable as "competing" with broadcast missed the real industry dynamic. In most cases, they became sibling divisions serving a common parent conglomerate.

These conglomerates took advantage of the greater capacity digital cable enabled by creating an abundance of new channels: ninety-four new channels launched in 1997 followed by one hundred in 1998![1] Many of the new channels didn't provide new programming so much as a secondary place to watch programs that already existed. For example, Lifetime expanded from its core channel and introduced Lifetime Movie Network (LMN) to create a distinct destination for its library of films that proved the most popular programming of the original network. This was valuable in a pre-DVR, pre-VOD (video-on-demand) era when the now-obsolete video cassette recorder (VCR) allowed time-shifting but was too unwieldy for regular use.

Subscribers' tentative shift to digital cable—which often came with added fees—meant there was a limited audience for the avalanche of channels that launched in the late 1990s. Channels that reached few homes earned little from cable service providers and had similarly limited advertising revenue potential. Channel owners aggressively launched new channels even before they could reach audiences to ensure a place among cable systems' offerings when audiences did arrive.

The transition to new competitive norms consequently developed slowly, but it was clear that bigger change was looming. The increased capacity of digital cable changed cable's competitive conditions. In an environment of a few dozen channels, simply offering something different from broadcasters—an old show from a broadcast network or a decade-old film—was an adequate strategy to draw enough viewers for a viable business.

But a 200- or 300-channel environment required a different strategy. In an era of digital cable that offered hundreds of channels, there were too many channels to "graze" or channel surf. Digital services also introduced electronic program guides that allowed viewers to quickly survey viewing options.

The coming surplus of channels made it newly important for cable channels to develop a distinct identity. A clear identity—or brand—would help channels stand out among the growing abundance of competitors. Viewers would first check channels they knew to regularly offer programs of interest. Channels such as CNN, ESPN, Nickelodeon, and MTV had a built-in brand because they were organized around a specific type of content.

But for others, developing a brand was complicated. Many other "general interest" cable channels existed—USA, TBS, TNT—and had succeeded to this point by featuring a schedule of programs previously developed for broadcast networks. As table 4.1 indicates, the most-watched cable programs in any week in the mid-1990s featured several sporting events—particularly WWE wrestling; movies—often five years after their release in theaters; and *Rugrats* and other Nickelodeon kids' cartoons. Series that did infiltrate most-watched lists were typically those important in subcultures such as MTV's *The Real World* (1992–) and *Beavis and Butthead* (1993–1997; 2011). Such series may have been hits with niche audience segments, but most weren't viewed by wider audiences or seen as meaningfully competing with broadcasters.

The reason U.S. television changed so considerably has to do with this shifting competitive dynamic. Creating original programming was a good way to build a channel's brand. And the combination of a clear brand and original programming was valuable to all the ways cable channels earn revenue. A clear brand makes a channel a reliable destination (which drives more viewers and thus advertising revenue), while original programming also could help channels negotiate higher fees from cable service providers. Channels that offered programs that viewers would truly miss were

Table 4.1
Top 30 cable programs, July 1997

Program	Channel	Households (000)
1. *World Wrestling*	USA	2,915
2. Movie: *Rough Riders Part 2*	TNT	2,882
3. *World Championship Wrestling*	TNT	2,870
4. *World Championship Wrestling*	TNT	2,455
5. *Rugrats*	Nickelodeon	2,433
6. Movie: *When the Bough Breaks*	Lifetime	2,309
7. *Maurice Sendak's Little Bear*	Nickelodeon	2,289
8. *Maurice Sendak's Little Bear*	Nickelodeon	2,265
9. *Hey Arnold*	Nickelodeon	2,215
10. *Blues Clues*	Nickelodeon	2,205
11. *Rugrats*	Nickelodeon	2,172
12. *Angry Beavers*	Nickelodeon	2,147
13. *Rugrats*	Nickelodeon	2,143
14. *Busy World of Richard Scarry*	Nickelodeon	2,137
15. *Real World VI*	MTV	2,081
16. Movie: *Godfather*	USA	2,131
17. *Blues Clues*	Nickelodeon	2,121
17. Movie: *Godfather II*	USA	2,121
19. *Busy World of Richard Scarry*	Nickelodeon	2,114
20. *Jim Henson's Muppet Babies*	Nickelodeon	2,109
20. *Maurice Sendak's Little Bear*	Nickelodeon	2,109
22. *Busy World of Richard Scarry*	Nickelodeon	2,107
23. Movie: *Godfather Saga*	USA	2,096
24. *Blues Clues*	Nickelodeon	2,094
25. *Blues Clues*	Nickelodeon	2,043
26. *Maurice Sendak's Little Bear*	Nickelodeon	2,031
27. *Allegra's Window*	Nickelodeon	2,024
28. *Rugrats*	Nickelodeon	2,013
29. *Rugrats*	Nickelodeon	2,010
30. *Gullah Gullah Island*	Nickelodeon	1,996

**Source:* Donna Petrozzello, "Basic Cable Ratings Up in July," *Broadcasting & Cable*, August 4, 1997, p. 49.

By far, most of what Americans watched on cable in 1997 was found on Nickelodeon. All but one series ranked 30 to 50 also aired on Nickelodeon.

in a stronger negotiating position for those fees than those channels that merely offered "something else" to watch.

General-interest channels such as TNT, USA, TBS, and eventually FX and AMC, sought—and drew—larger audiences, which provided the budgets to engage in the most ambitious experiments with original series development. Because of their lack of niche-specificity, they were more like—and thus could more effectively challenge—broadcast television. It seemed most unlikely in the 1990s, but within a decade, general-interest channels developed programs that rivaled those of broadcast networks.

5 Cable's Image Problem

The birth of many cable channels can be traced back as far as the 1970s, but even though the channel names may be familiar, the content of these channels and their availability was very different than it became by the early 2000s. Channels were not reliably found on every cable system. This inconsistent carriage led advertisers to regard cable channels as better suited for local and regional advertising, which limited advertising revenue.

In addition to concerns about the varied availability of channels across the country, advertisers also were skeptical about the cable audience. Because cable channels were perceived as offering a lesser form of programming, those who chose to view it—even though they paid for a cable subscription—were consequently viewed with some suspicion. Many advertisers simply dismissed cable as a wilderness of children, teens, and those with nothing better to do than watch television all day.

Concerns about reach and perceptions of cable as an inferior form led cable's advertising rates to be significantly lower than broadcast. This weaker revenue stream made it difficult for cable channels to amass funds for original programming that might shift those perceptions.[1] Advertisers paid roughly half as much to reach 1,000 households on cable as they paid for the same audience on broadcast networks in 1996 ($5.95 versus $10.40 for households in prime time). To be clear, these rates are for the *same* number of viewers. This disparity lessened some by 2014, but remained significant.[2]

To the frustration of cable programmers, some of the most viewed cable programming had difficulty attracting advertisers. In the early 1990s, USA's weekly two-hour block of WWE wrestling regularly collected larger audiences than anything else on cable. However, many media buyers and advertisers perceived wrestling as tawdry and not something with which brands should be associated. So even when cable channels found substantial audiences, advertisers were not always forthcoming. The lower advertising rates and relatively low fees paid by cable systems proved a substantial impediment to amassing the budgets needed to create original series.

The challenge for cable in its struggle to be seen as a legitimate part of television wasn't simply a matter of launching channels and convincing advertisers to reconsider expectations of television buying built on broadcasting. It also required persuading viewers to reimagine their expectations of television. There was little approximating "appointment" television on cable at this time. Cable was where you flipped during a commercial or after deciding there was nothing to watch on broadcast networks. Few cable channels would have survived if they had been entirely dependent on advertising revenue, but cable channels also drew revenue from fees paid by multichannel service providers for each household that could receive the channel. These subscriber fees equaled the advertising revenue for large channels. Cable channels, though, still could hardly afford to develop the type of programming likely to attract audiences when being paid half as much for advertising.

Films were the first form of scripted, original programming cable channels developed. They quickly became plentiful, but producing original films was costly. Cable channels initially produced films with budgets of two to five million dollars, but they soon climbed to a norm of eight million. Special event films could top twenty million. Many of those event films were developed with the expectation that they could be sold in international markets so the cable channel typically didn't shoulder the cost of production alone.

On the other side of the financial equation, these original films also produced significantly higher advertising income. In 1997, TNT president Brad Siegel explained that original movies on TNT earned advertising rates of $13 to $16 per thousand viewers, which was double or more than the channel received on average.[3] Many channels were rewarded for the risk and effort involved in developing original films.

By the mid-1990s, original films were common on cable, and cable had largely wrested the form from broadcast networks. Cable could design movies targeting more narrow audience interests and tastes, and broadcast movies struggled to gather mass audiences once cable films grew common. Cable channels could also air a film multiple times and amortize costs across the time slots in a manner unavailable to broadcast networks. In the 1998–1999 season, A&E, USA, TNT, and Lifetime collectively aired seventy-five original films. Cable competition led broadcasters to eliminate many of their weekly movie-night time slots throughout the 1990s.[4]

Original programming of any kind was advantageous because it allowed channels to more directly define their identity. Acquired content—series and films developed for broadcast networks or theatrical release—typically had a more general brand and often was closely associated with its original source.[5]

Original films were less risky for channels than series because their costs were lower and because a failed movie had fairly narrow impact. Likewise, however, a successful original film produced limited benefit for the channel. Unlike a series that was likely to return an audience each week, original films required an exceptional amount of promotion for each new one. Producing original films also became less exceptional once more channels did so, which decreased the ability of original films to make a channel stand out from the competition. Innovative channels again sought new strategies for distinguishing themselves from the growing pack of competitors. Original series provided that innovation.

6 The Long Road to Original Cable Series

Why didn't cable channels create original programming until the mid-1990s? The short answer is money. Understanding the many different ways money was an issue requires a longer answer.

The key factors prohibiting development of original series at this point were interrelated: their cost, their risk, and the difficulty of marketing them. Some efforts toward original scripted series emerged in the early 1990s—WTBS even attempted a block of low-cost original comedies in the mid-1980s—but series developed for ad-supported cable were short lived.[1] The Family Channel experimented with the scripted family comedies such as *Big Brother Jake* (1990–1994) and *The Mighty Jungle* (1994), which were similar to broadcast series. These shows did not have a significant effect on the perception of the channel nor did they earn much notice by viewers. The series were unremarkable in terms of innovation; if anything, they were more conventional than broadcast comedies such as *The Cosby Show*, *Roseanne*, and *Murphy Brown*, which drew large audiences on broadcast networks at this time.

Niche interest cable channels first tested the waters of original series development with unscripted shows such as *The Real World* (1992) and *Road Rules* (1995) and animated series such as *The Ren and Stimpy Show* (1991–996), *Beavis and Butthead* (1993), and *Space Ghost Coast to Coast* (1993–2008). Nickelodeon was the clear pioneer—it had taped more than one thousand hours of programming by 1994.[2] There were also some early unscripted series—although beyond *The Real World* and *Road Rules*, original "reality" series were rare in the 1990s. Animal Planet began airing *The Crocodile Hunter* in 1997 and TLC launched *Trading Spaces* in 2000. Unscripted series had the advantage of being cheaper to produce; we will explore them in depth in chapter 14.

Series' risks derived not simply from the difficulty of predicting and developing successful programs, but from uncertainty about whether audiences would even watch a weekly series on a cable channel given perceptions of cable. Much of the broadcast networks' dominance derived from the habits and behaviors they had created in viewers. Audiences were accustomed to turning to broadcast networks every night and flipping through what was on. Cable channels weren't typically among those automatically considered. Viewing habits guaranteed broadcast networks a consistent level of viewership and a ready platform for promoting other programs on the network.

Cultivating viewers' expectation of finding something on cable would take time. Thus, it seemed risky to the executives at any cable channel to try to be the first. Even if cable channels sunk millions into developing series, the promotional budgets required so that audiences might actually find them were prohibitive. No one yet conceived of cable's ability to target more particular tastes or audiences as a strategy, let alone a competitive advantage.[3]

The emergence of original series appeared haphazard and unclear to viewers of the late 1990s. Looking back, two preliminary strategies gave rise to the one that would—in time—come to seem obvious and prove most successful. Finding these strategies, however, came only after failed experiments with others.

With few exceptions, the first original scripted cable series—particularly those on advertiser-supported cable—are best categorized as *low-budget imitations* of broadcast series. USA's *Pacific Blue* (1996–2000) was *Baywatch* but with bike cops, *Big Brother Jake* (1990–1994) was a conventional family sitcom set in a foster home, and *USA High* (1997–1999) was a teen sitcom in the vein of *Saved by the Bell*. In this phase, cable channels simply tried to compete directly with broadcasters, but they were disadvantaged by their more limited budgets, resulting in largely forgettable programming.

The failure of these series led the more innovative cable channels to find ways to increase budgets. *Coproductions* then emerged as a second strategy as cable channels pooled their budgets with partners. The first successful experiment involved a coproduction between USA and CBS to create *Silk Stalkings*, which followed the exploits of a detective team assigned to solve crimes of passion in glamorous Palm Beach. The second experiment, *La Femme Nikita*—a coproduction between USA and partners in Canada, Germany, and France—was the first original cable series offered by advertiser-supported channels that made a dent in expectations of cable.

It is important to note that among cable channels there exist two sectors with different business models. Advertiser-supported cable—the focus thus far—had two revenue streams that derived from selling audiences to advertisers and the fees paid by the cable service providers to include the channel on their service. Subscriber-funded cable—channels such as HBO and Showtime—included no advertising. These channels had to offer a service with such value that people would willingly pay for them.

Subscriber-funded cable channels largely skipped the stages of low-budget imitation and coproduction. They experimented with developing programs clearly distinctive from broadcast norms from the outset. Subscriber-funded channels also began producing original series earlier. HBO produced a comedy about a football team, *1st and Ten* (1984–1991) in the mid-1980s and followed it with *Dream On* (1990–1996), and *The Larry Sanders Show* (1992–1998). Similarly, Showtime had *It's Garry Shandling's Show* (1986–1990), *Red Shoe Diaries* (1992–1997), and *The Outer Limits* (1995–2002).[4]

There were other series, but many were uninspired, being distinctive only for their inclusion of gratuitous amounts of what the industry vernacular described as "T&A." Both *It's Garry Shandling's Show* and *The Larry Sanders Show* drew critical attention and were popular with a segment of the already niche subscription cable audience. These isolated shows seemed aberrations more than an emerging strategy.

The 1996–1997 television season was the first year cable channels' original programming efforts gained substantial notice. Trade magazine *Broadcasting & Cable* produced a special report on cable originals in February 1996 that explored these experiments as significant and innovative. This was not the first year cable channels tried original scripted series, but these series were the first that suggested such efforts were more than novelties and might redefine expectations of cable programming.[5]

The most exceptional early experiment—and the one that best foreshadows what cable originals would become—was a show few recall: AMC's *Remember WENN* (1996–1998). At the time it aired, AMC was a classic movie channel that reached 64 million homes. *Remember WENN*—a dramedy long

Table 6.1
Cable original scripted series on subscriber-funded channels before 1996

Not Necessarily the News	HBO	1983–1990
1st & Ten	HBO	1984–1991
It's Garry Shandling's Show	Showtime	1986–1990
Tales from the Crypt	HBO	1989–1996
Dream On	HBO	1990–1996
The Larry Sanders Show	HBO	1992–1998
Red Shoe Diaries	Showtime	1992–1997
The Outer Limits	Showtime	1995–2002
Mr. Show with Bob & David	HBO	1995–1998
Tracey Takes On ...	HBO	1996–1999

before the term was common—was set in 1930s Pittsburgh and told the stories of the employees of a radio station that was still performing live radio drama and comedy. Devoid of a laugh track or commercials, the scripts—written by dramatist Rupert Holmes—were composed as one-act plays.[6] AMC followed *Remember WENN* with *The Lot* in 1999, which was set in 1930s Hollywood and blended fiction and reality to tell behind-the-scenes tales of those in the early film business.

Both series were short lived, but they indicated significant aspiration, demonstrated an understanding of how original programming could establish and promote a channel brand, and were clearly an effort to create programs truly distinct from broadcast fare. AMC was an especially niche channel at that time, and it was too soon for innovative programs to help broaden its audience beyond those interested in viewing classic films.

To dive more deeply into the challenges cable channels faced in creating original programming, we will explore case studies of two of the first successful series in the next two chapters. Choosing a starting point in any history is fraught and contentious. There is not an unassailable logic that allows declaring *La Femme Nikita* and *OZ* as important first milestones in original cable series rather than *Remember WENN, Babylon 5, Pacific Blue,* or *The Larry Sanders Show*. This carefully considered decision ultimately results from matters of timing as well as judgments about different series' and channels' aspirations and accomplishments.

Beginning with the stories of these less-remembered series is strategic. Both series faced many challenges related to being unprecedented that neither fully overcame. This makes for a much richer first set of stories about qualified success than if this journey began with the more obvious accomplishments of *The Sopranos, The Shield,* or *Monk*. Considering a series from an advertiser-supported and a subscriber-funded channel also illustrates how these different revenue models affect the programming strategies of these different cable sectors.

7 Cable's First Antihero

In 1996, USA was the second most-watched cable channel in prime time behind general-interest competitor TNT. Its strongest ratings were earned by its Tuesday evening WWE wrestling, Wednesday evening boxing, and series acquired after they had aired on broadcast networks that were scattered throughout the day. Wrestling often drew the largest audience for all of cable in the week. But advertisers' concerns about associating their products with wrestling made it a dubious tent pole for USA to build a channel identity around.

USA experimented aggressively with developing original series. As noted, it achieved limited early success with a series coproduced with CBS in the early 1990s, *Silk Stalkings* (1991–1999). CBS first aired each episode as part of its late-night "Crimetime after Prime Time" schedule. USA then re-aired it in prime time later in the week. A conventional detective show from prolific producer Stephen Cannell, *Silk Stalkings* had the look of a low-budget series. It ignored much of the innovation evident in broadcast detective shows such as *Miami Vice* or even Cannell's own *Wiseguy*. Two years in, CBS gave up the late-night block when it prepared to launch *The Late Show with David Letterman*. USA continued the series, which drew modest audiences but little prestige or notice on its own.

USA's next original drama followed the aspiration of *Silk Stalkings*. *Pacific Blue* (1996–2000) was formulaic, yet drew some viewers and was sold to stations in several international markets. It didn't attract critical interest and did little to disrupt broadcast dominance or bring recognition to USA. More than anything, it reaffirmed cable's reputation as a site of mediocre programming.[1]

Between the launch of *Silk Stalkings* and *Pacific Blue*, USA attempted to develop an evening of original comedy series: the animated *Duckman*

(1994–1997), *Weird Science* (1994–1998), *Campus Cops* (1995–1996), and a few others that aired briefly and left little trace. Ironically, *Duckman* offered what was probably the best glimpse of the future of cable programming. The show was ahead of its time in terms of establishing a clear comedic voice and was as likely to turn viewers away as passionately engage those who shared its sense of humor. Based on a comic, *Duckman* was irreverent and adult, similar to *The Critic* (ABC, Fox, Comedy Central, 1994–1995) and *Dr. Katz* (Comedy Central, 1995–2002), which likewise appeared before *South Park* truly established the outer limits of popular profane animation in 1997. USA's comedies were clearly distinctive in tone, but none was successful enough to draw substantial attention or viewers to the channel.

USA's first shows looked familiar and cheap. The dramas reproduced tired genres and character types and had an undeniable low-budget look. They pushed boundaries only in terms of being a little racier than broadcast fare—illustrated well by the decision to confine *Silk Stalkings* to broadcast late night. Even broadcast series had become more innovative through the early 1990s: *ER, Homicide: Life on the Street, Law and Order,* and *NYPD Blue* pushed the well-worn forms of medical and detective series in new creative directions. In comparison, cable's experiments with original series seemed dated and consequently were unlikely to draw audiences away from broadcast networks. Even the old broadcast series that filled most cable channels' schedules provided better entertainment.

But USA's *La Femme Nikita* (1997–2001) began to change all that. The series was based on the 1990s French action-thriller film *Nikita,* directed by Luc Besson. It tells the story of a woman blackmailed into joining an elite counterterrorism organization, where she must work as an assassin. The series was more psychologically layered than most action series because the mission-of-the-week plot was augmented by an ongoing story of Nikita's struggle to maintain her moral code. The series also developed slow-burning romantic interest between Nikita and another agent, Michael, to add other emotional layers and interest. Of other well-known series, it was most similar to ABC's *Alias,* which debuted four years after *La Femme Nikita*.

It took nearly seven years to bring *La Femme Nikita* to the screen and little about its story could be called routine.[2] It likely never would have been created without the tenacity of Marla Ginsberg—an executive at the French studio Gaumont that produced the 1990 film that originated the series' concept. The struggle to develop *La Femme Nikita* reveals the many ways the television industry was built to reproduce the business established by broadcast television.

As soon as Ginsberg saw the film, she knew that it would be a great concept for television. She quickly learned, however, that Gaumont had included the television adaptation rights along with the film rights sold to Warner Bros., which remade the film in 1993 under the title *Point of No Return*. The intrepid Ginsberg went to Warner Bros. to convince them of the potential of a *La Femme Nikita* series. The studio didn't share her conviction, so Ginsberg began pitching the idea to different channels in the hope that securing a potential buyer would persuade Warner Bros. to develop the series.

Few saw the potential Ginsberg did. Series centering on a female lead character remained uncommon in the early 1990s, and seemed a risky gamble. The series wasn't like anything on the air at the time. In some ways it was most like 1970s series such as *Charlie's Angels* and *Wonder Woman* with its mission-of-the-week storytelling, but it was dark, very dark, in tone. Often shows that are unlike anything on television—those describable as "distinctive"—were dismissed as too improbable to succeed. But distinctive was about to become a key strategy.

Ginsberg nearly secured a deal for the series from Lifetime—then marketed as "Television for Women"—but when that failed, she pitched the series a second time at USA.[3] The channel had a new executive team in place, and Rod Perth, president of the network, was looking for a series that would brand USA as a destination for original content. Besson's *La Femme Nikita* also happened to be one of his favorite films.[4]

It was still a long road for the series even after development began at USA. Having a cop-killer-turned-assassin lead character remained too dark a concept for USA. CEO Kay Koplovitz was particularly worried about whether advertisers would support such a show (a prescient foreshadowing of what would happen when FX attempted the edgy *The Shield* seven years later). The writers tried to aid Nikita's likability by changing the character from a cop killer to a woman framed for murder who is then forced into a counterterrorism agency with unclear alliances.

USA continued to worry about the tone and then almost gave up on the series when it seemed impossible to find an actress who could embody the complexity of the lead role. The key challenge here involved identifying a skilled actress who would not only work on television, but would also work on a series produced for a cable channel that was likely to have a bare-bones budget. Over two hundred actresses were considered before auditions, and more than fifty women read for the role. A little-known Australian model starting an acting career named Peta Wilson was the favorite of both the studio and USA. Although she had reservations, Wilson was willing to sign on for a role in a series made for a cable channel.[5]

After these and various other near-catastrophic challenges, *La Femme Nikita* was set for a January 1997 debut on USA. Broadcast networks schedule a lot of reruns in January, and this allowed its premiere to face less competition. USA also drew some its largest audiences to the movies scheduled for Christmas week, which provided the channel many opportunities to promote the new series.

La Femme Nikita was not a clear hit from the start. About two million households watched the first episode, a disappointing result given the promotional effort and budget USA had given the series. USA soon moved *La Femme Nikita* from Monday night—the home of highly viewed WWE wrestling—and built a programming block branded as "Sunday Night Heat," with *Pacific Blue* leading into the series and *Silk Stalkings* following it. Audiences increased. The second season debut in January 1998 drew the highest viewership USA had achieved for an original scripted series, a record that stood for five years.

Although "successful," *La Femme Nikita* wasn't the game changer that some later original cable series were. USA's executive team understood how the business of television was about to change, but there is only so much a single series could do. When *La Femme Nikita* debuted, USA's most popular series included WWE wrestling, boxing, and *Murder, She Wrote*—a mystery series that featured Angela Lansbury as a sexagenarian detective, which had been developed for CBS in 1984. Even as the highest-rated original drama on cable, it was difficult to use *La Femme Nikita* to promote USA's other programs because there was little that connected these series. In these earliest days of developing brands for general-interest cable channels, Perth's strategy of defining USA as a site for original programs exclusive to USA was a step forward, but not a big enough one.[6]

A shift in corporate ownership at USA resulted in the ouster of Perth and much of his programming team in 1998. The new executives offered *La Femme Nikita* little support. It began receiving less promotion, and the channel executives asked its producers to write episodes featuring WWE wrestlers in a cross-promotion attempt (although Perth's team learned early on that there was little crossover between these audiences). The channel's next slate of original programming that debuted in 1999 and 2000 failed quickly (*G vs. E, Cover Me, The Huntress, The War Next Door, Manhattan, AZ*). Following these failures, USA began to regard original series as an "elective" rather than a core part of its strategy.[7] Not until the debuts of *The Dead Zone* and *Monk* in 2002, under yet another new executive team, would USA achieve another original series success.

La Femme Nikita met a premature end in 2001 partially because it was orphaned by its executive team, but mostly as a result of the perceived limitations of the business model for original scripted series at this time.[8] USA paid $350,000–400,000 per episode to Warner Bros. for *La Femme Nikita*, although the series budget was $900,000 per episode. The series never could have been made without a cofinancing deal in which partners in Germany, France, and Canada helped finance the budget in return for the right to air the series in their countries.

When *La Femme Nikita* debuted, there was no track record for series created for a cable channel being sold repeatedly in the United States and internationally—which is how television series earn most of their revenue. Warner Bros. reasonably expected that there would be few opportunities to earn additional revenue from the series. Once the series had produced enough episodes to sell in other markets, the studio saw no value in continuing to produce new episodes. So, somewhat curiously, Warner Bros. effectively canceled the series.

La Femme Nikita's cancellation was announced just as the series was shooting its final episode, which wasn't intended as a conclusion, but a cliffhanger that would significantly reboot the series. An extensive fan campaign to bring the series back combined with challenging times at USA—it had lost its rights to wrestling—and led the channel to develop a deal for a fifth season of eight episodes at a price that made it worthwhile to Warner Bros. to properly conclude the series.

The story of *La Femme Nikita* reveals a lot about the challenges of inventing the original series business for cable. Its struggles resulted both from trying to do something unprecedented and attempting a series so unconventional.[9] *La Femme Nikita* was uncommon in its tone, in its genre, and in featuring a central female character. From a vantage point twenty years after its debut, this distinctiveness isn't as obvious. But all three of these characteristics marked it as distinctive at the time.

The series was just as exceptional in what it revealed of the business practices of television at that point. In the late 1990s, the idea of creating a series for a cable channel that might be so good that other channels and broadcast networks would want to buy the rights to re-air the series was about as conceivable as claiming the Earth was round in the 1400s. Although viewers were captivated by how much television began to change as cable began producing original series, behind the scenes, the business of television was being just as radically rewritten.[10]

8 *OZ* Locks Up Cable's New Strategy

Just six months after the debut of *La Femme Nikita*, HBO launched *OZ* (1997–2003), a series rarely afforded its proper due for its role in preparing the ground from which milestones such as *The Sopranos* and *Sex and the City* grew. Although HBO had aired distinctive original comedy series such as *The Larry Sanders Show*, no series before *OZ* so radically pushed the boundaries of U.S. television.[1]

There had always been some distinctive programming in the United States. But before a competitive environment of hundreds of channels became established, those distinctive programs were rarely commercially successful. There was little room in the broadcast television business to make shows small audiences found interesting because the business model measured success by gathering the most eyeballs.

OZ also had a key advantage over *La Femme Nikita* in the quest to make distinctive television. A subscriber-funded business, like HBO's, measured success differently than USA, which relied significantly on advertising. HBO earns its revenue by creating programs viewers find so compelling that they are willing to pay for the service. It measures success in the number of subscribers who find enough value to warrant paying for service each month. Advertisers aren't a consideration. In contrast, advertiser-supported television succeeds when it can attract a big audience of viewers advertisers desire to reach.

Shows that become great milestones in television history rarely start out with such potential being apparent. In most cases, the shows that retrospectively "changed everything" were just one risk-averse executive away from never being made and could have just as easily gone into the heap of long-forgotten failures. *OZ* represented a stunning divergence from what

American audiences expected from television. Its role in television history was consequently most unanticipated.[2]

HBO may have been the first cable channel when it began transmitting in 1972, but as of 1997 it had spent its quarter-century of existence trying to be as little like television as possible. In an era before viewers could rent films on videocassette tapes, HBO was a service that brought movies, stand-up comedy, and some special sports events into American living rooms. HBO began producing original films in the early 1980s to further cement its connection to film. This was a valuable strategy in an era in which access to films was scarce, but by the 1990s, viewers had other options.

By focusing so much on not being television, HBO missed out on one of the most valuable attributes of the business of television—a regular schedule. Neither HBO films, theatrical movies, boxing matches, nor comedy specials took advantage of a key strategy that broadcast networks had capitalized upon since the 1960s. Broadcast scheduling emphasized regularity—same program, same time slot, every week. This regularity reduced promotion and marketing requirements exceptionally. If subscribers came to value HBO not just for special, one-off events but as a provider of exclusive, regular programming, then that would contribute a new strategy in reducing *churn*—the term for the number of people who cancel subscriptions each month.

Internal research and focus groups in the early 1990s revealed that subscribers were frustrated and confused by HBO's lack of a regular schedule. In response, HBO instituted the scheduling strategy of debuting a new theatrical film every Saturday at 8:00 p.m. and found that satisfaction with the service among subscribers who knew of the regular debuts increased 20 to 30 percent.[3]

HBO strategy really began to change when Jeff Bewkes took over as CEO in 1995. Bewkes doubled the original programming budget between 1995 and 1998 and mandated that Chris Albrecht, HBO's head of original programming and independent productions, develop original series.[4] But what type of series should the channel develop?

Much of what would be solidified through its original series of the late 1990s as the "HBO brand" was established by the original movies and miniseries that HBO produced and was lauded for over a decade before the appearance of its first series. HBO films were ambitious and relevant, and they often explored topics unlikely to be addressed in films released in theaters.[5]

Establishing the HBO brand through the creation of new series required developing shows unlike those on broadcast television. At the time, there was much that U.S. broadcast television wasn't doing and had never tried. There was no independent television sector outside of the mainstream that suggested ways to push boundaries. And the U.S. market seemed determined not to take lessons required from international markets, where several public service broadcasters developed creative series far afield of U.S. broadcast norms.

One strategy HBO identified during its experiments with original comedy series during the 1990s was that of developing different programs for different subscribers rather than trying to make shows that would appeal to all. For example, *The Larry Sanders Show* closely matched the savvy, unconventional identity HBO perpetuated with its films. *The Larry Sanders Show* was as smart and edgy as U.S. television could be at the time, but HBO also programmed *Arli$$*, a comedy about a sports agent that aspired to provide a behind-the-scenes look at sports the way *Sanders* did with the business of celebrity talk shows. Although the aspiration was similar, the results differed considerably. Critics who fawned over *Sanders* critiqued *Arli$$* as one of the worst shows on television. Despite that, and even though it deviated significantly from the brand HBO sought to cultivate, *Arli$$* was one of the most popular series on the channel. HBO kept *Arli$$* on for seven seasons because it added value to HBO subscribers who weren't seeking the critically acclaimed, buzz-worthy content commonly associated with its brand.[6]

Buzz was—and remains—central to HBO's strategy. Bill Mesce, who spent twenty-seven years working for HBO, explained that one of the ways HBO evaluated the success of series was with a metric that accounted for the dollar value of the visibility that HBO received from press coverage.[7] Every critics' column or cultural trend story that talked about HBO content served as free promotion for the service.

One of the best ways to generate buzz is to be different. When HBO set out to create its first original drama, its only aim was a show that would stand out. How *OZ* came to be that first drama is a story of chance as much as anything. Anne Thomopolous was the HBO executive charged with developing HBO's first drama; she knew it had to be different than the dramas airing on broadcast networks. HBO's first effort wouldn't be a subtle variation of a familiar show like *Silk Stalkings* or *Pacific Blue*.

HBO did not set out to create a prison drama. In conversation, Thomopolous's boss, Chris Albrecht, mentioned to studio executive Rob Kenneally that HBO's documentaries about life in prison had been particularly successful.[8] Kenneally happened to be one of the few people Tom Fontana had told about his interest in writing a series that explored life in prison—certainly unlike anything on a U.S. broadcast network.

In 1996, Fontana was writing and producing *Homicide: Life on the Streets* for NBC. *Homicide* was one of several shows pushing the boundaries of broadcast conventions of that time with innovative camera work, uncommon police stories, and richly crafted characters. After telling so many stories that ended with someone being sent to jail, Fontana thought it would be interesting to tell stories about what happens there.

Almost everyone he told about his idea dismissed it. Prison stories would be too gritty for broadcast television and likely to make advertisers uneasy. Fontana had pitched more "gentle variations of the idea—a juvenile detention center, a 'Club Fed' for white-collar criminals."[9] After talking with Albrecht, Kenneally called Fontana, and Fontana and producing partner Barry Levinson were soon pitching Albrecht and Thomopoulos. Albrecht recalls the situation simply as a decision to give Fontana one million dollars to make a twenty-minute pilot—not the typical process for developing a television show. When the executives saw the edgy result, they knew it would stand out.[10]

And stand out it did. *OZ* broke nearly every rule—from unexpectedly killing off protagonists, to having a narrator character that speaks directly to the audience, to populating an entire series with unlikable characters. Even these innovations in storytelling form don't address *OZ*'s unflinching—yet also not exploitative—inclusion of all manner of sex, violence, and language. *OZ* showed things never previously aired on television. Much of it couldn't even be found in movies.

OZ took place nearly entirely in the prison. It mainly told the stories of the inmates and the complexity of the social and power structures that developed from violent criminals living together. The series often revealed, by means of flashbacks to their crimes, how the prisoners became incarcerated. *OZ* didn't feature the story of a particular character, but had four inmates and three prison workers who spanned the full series. Each season featured a mix of storylines about as many as eight prisoners, and sometimes of prison guards as well, over multiple episodes. Few inmates survived all six seasons.

Just as HBO miniseries and movies were provocative, so was *OZ*—not in a tawdry way, but in the sense that it provoked an audience to think about the nature of prison, of good and evil, and even of storytelling strategies and filmmaking conventions. Unlike many of the series that followed it, *OZ* was not a reimagined version of a known genre, nor was it drawn from existing intellectual property, such as a book or film. *OZ* could be argued as the most original series HBO would produce.[11]

Although HBO's subscriber funding encouraged a strategy of distinction, a lot of the distinctiveness of *OZ* resulted from its unusual development process and the creative freedom HBO afforded Fontana. These are creative management strategies that are separate from the difference of its revenue model.[12] Rather than first asking Fontana for a script they could review and offer feedback on, HBO simply gave him the one million dollars to put his vision on film. That meant they received his own vision, not his vision altered by suggestions from channel executives after reading a script, which is the typical process. Fontana's aspirations were so far from television norms of the time that they couldn't reasonably be imagined. HBO's willingness to let Fontana illustrate the feasibility of the idea by putting it on film before disregarding it was crucial. We can only speculate about how many amazing series missed being created because they were inconceivable in a pitch or on paper.

Although creative freedom would soon become characteristic of the HBO brand, there is no reason related to HBO's subscriber funding for this to be so. Subscriber funding encouraged distinction and creating programs people would pay for rather than those that attracted the largest audience.[13] Creative freedom was simply a managerial strategy. Faced with launching its first drama, Albrecht acknowledged that his creative background was in comedy, not drama, and he gave Fontana free reign. Thomopolous recalls encouraging Fontana to go back to scripts and make them edgier and grittier, and push the boundaries farther—the opposite of the common experience at broadcast networks.

Fontana recalled receiving only one note from HBO. In a flashback scene that depicts one of the inmates murdering an entire family, the camera lingered—long enough to make an HBO executive uncomfortable—on a shot from the inmate's point-of-view as he prepared to shoot the children.[14] Merely speeding up the shot addressed the problem without making Fontana feel his vision had been compromised, and it maintained the writer's creative freedom. Fontana recognized what a rare gift he had been given and he also felt a duty to those creators who might follow him to be sure that it was an opportunity HBO didn't come to regret.[15]

Of course HBO's creative management strategy could be deemed successful only if *OZ* achieved what HBO desired for the series to accomplish. In testing the waters of original drama production, HBO was less concerned with how many watched or even an immediate uptick in subscribers. Albrecht sought for HBO series to drive buzz and increase the frequency of mention of HBO in the media. *OZ* needed to establish HBO as a creator of important shows.[16] It did this, perhaps not as noisily as subsequent series, and it certainly drew attention.

OZ was not as successful as some of the series that followed it in terms of the number of viewers who tuned in. It averaged four million viewers for its Sunday premiere episodes by 2002.[17] It also debuted before selling series on DVD emerged as a valuable revenue stream. DVD releases were especially valuable for HBO because they were the only way people who didn't subscribe could see its series. *OZ* was also sold in at least twenty-four other countries, which provided further revenue for the series. The graphic content likely curbed the markets and earnings available. As with *La Femme Nikita*, international buyers had no knowledge about the performance of dramas created for a U.S. subscriber-funded channel, which likely led to hesitation about buying it.

HBO's original series development after *OZ*'s debut is well known. First, *Sex and the City* (1998–2004) and then *The Sopranos* (1999–2007) set the nation abuzz to an extent that few details need to be repeated here. Development continued slowly but steadily with the HBO brand being defined primarily by the originality and distinctiveness of subsequent series: *Curb Your Enthusiasm* (2000–2011), *Six Feet Under* (2001–2005), and *The Wire* (2002–2008). The channel added just one new series a year in this time frame (excepting the two-season stumble of *The Mind of the Married Man* [2001–2002]), but drew considerable attention.

By the early 2000s, HBO had not significantly increased its subscribers, but the wider culture couldn't help but notice something was going on. There was a lot of talk about HBO and its shows as it began dominating awards, although talk of a new "golden age" of television had not yet begun. Ironically, HBO advertised itself with the slogan "It's Not TV, It's HBO" at the same time it was developing original series to build regularity and consistent viewing. This slogan matched how HBO had long seen its identity even though it now embraced a strategy of being more like television. Its aim may have been distinction—including being seen as distinct from "TV"—at least from television's broadcast past. But its innovation had enormous implications for the future of television.[18]

9 Seeds of Transformation

This story of cable channels' struggle to develop distinctive series isn't meant as a rebuke to the shows produced by broadcasters or the audiences that loved them. Understanding the process of television's transformation requires identifying connections between shifting competitive conditions and how those shifts encouraged changes in television storytelling.

It wasn't the case that great television storytellers telling distinctive stories didn't exist before the late 1990s. Rather, the strictly observed practices for creating television designed to gather a mass audience inhibited opportunity. Most television looked a certain way because of those norms and practices. There were exceptions—but they were precisely that. Cable channels' need to stand out upon the arrival of digital cable and satellite, their dual revenue streams, and their hundred-channel environment made possible new norms, practices, and measures of success that led to other types of programming. As these early cases illustrate, this was a long and challenging process of change.

The arrival of cable as a site for original scripted series is important because it creates a sector of television storytelling governed by different business practices and constraints. The series created in the late 1990s by HBO enabled audiences and those in the television industry to imagine television differently. The business and management practices of broadcast television had narrowed innovation. But these first original scripted cable series broadened perceptions of what could be done on television. Of course, the possibilities they introduced were inextricably tied to the very different goals and expectations of what could be considered successful. The HBO shows helped reveal that it wasn't the *medium* of television that had a narrow palette of creative possibility and that a sizable segment of the U.S. audience hungered for more ambitious storytelling.

Although this story argues that cable was the source of transformation, it wasn't that broadcasters were blind to innovation. In many regards, broadcasters' business model handcuffed them. Broadcast television developed unconventional content before cable competition and especially after. *The X-Files* (1993–2002), *Homicide: Life on the Streets* (NBC, 1993–1999), *NYPD Blue* (ABC, 1993–2005), and *ER* (NBC, 1994–2009) proved commercially successful and reinvigorated established genres. Broadcasters also attempted greater experimentation with series such as *Wiseguy* (CBS, 1987–1990), *Twin Peaks* (ABC, 1990–1991), and *Murder One* (1995–1997), but they failed commercially. Broadcast networks responded to the critical praise of HBO's late 1990s dramas with series that employed a variety of unconventional storytelling tools and strategies such as *Alias* (ABC, 2001–2006), *24* (Fox, 2001–2010), *Push, Nevada* (ABC, 2002), and *Boomtown* (NBC, 2002–2003) even before advertiser-supported cable could reproduce *La Femme Nikita*'s success. These series were limited successes in terms of the metric of audience size, the only important metric for broadcasters. Relying solely on advertiser dollars pushed broadcast networks toward mass hits, not niche distinction.[1]

It is helpful to view the original series developed by broadcast networks and cable channels as engaged in dialogue rather than a battle. The large media content companies owned both broadcast networks and cable channels. From a business perspective they were sibling divisions that corporate parents used to diversify risk as the television business evolved. Although the two industry sectors competed for viewers' attention—especially in the years in which a schedule still dominated access to programs—they did so with disparate opportunities created by the differences in their businesses and measures of success. Determining a "winner" of this competition was much less interesting than exploring how the shows of this era attempted to leverage the conditions of being either a cable or broadcast series while still clearly being inspired by the efforts and accomplishments of the other.

10 The Death of Television! It's a Golden Age of Television!

By the early 2000s, two nearly opposite ideas about television competed. One proclaimed the death of television; the other argued that television was in the midst of a new golden age. Both claims were hyperbolic and ignored all but the industry's most recent past. There were kernels of truth in both claims, but their contradiction led to confusion that mostly affirmed the sense that the business of television was in peril.

Part of the contradiction came simply from confusion about what television was. Often, what was intended by claims of its death was really just an assertion about the end of the domination by broadcast networks. The praise and attention levied on HBO's original series could often seem like a dismissal of all other television. Significantly, no cable channel other than HBO created a noisy original series success until 2002, so it was unclear whether "cable" was really a contender, or just HBO.

There was also a sense that coming technological disruption was imminent, but the nature of that disruption was unclear. In the early 2000s, streaming video—such that it existed—was nothing like what became the norm around 2010. Images loaded slowly, would play with significant pixilation even when complete, and often inexplicably stopped in midstream. The experience was awful, and there was minimal content worth watching. The media companies that could have provided programming worth viewing feared the threat presented by internet-distributed television and made little other than promotional material and "extras"—such as behind-the-scenes videos—available.[1]

Industry executives who didn't embrace strategies that pivoted their companies to take advantage of this new technology should be forgiven based on the state of early internet distribution. It was impossible to imagine how quickly the technology would improve. Likewise, it was highly unlikely that a savvy distributor with no legacy television business would emerge (Netflix), or that portable and mobile screens would develop at just that moment as well. It was clear in the early 2000s that significant change was on its way if one looked at the similar disruption in the print and music industries. Yet, it continued to be impossible to imagine television outside of its broadcast norms.

It is helpful to remember what "new media" entailed in the early 2000s. Both the recording and newspaper industries provided fearsome, cautionary tales for the video industries. Television and film were spared disruption by internet distribution until 2010 because the file size of video and the state of compression technology and processing weren't yet advanced enough to allow video to zip across the internet as quickly as print and audio files. Internet distribution of all media was still new in the early 2000s. The first iPods came to market in 2001 and iTunes launched in 2003. For the most part, the music industry was still in free fall from illegal downloading.

Most newspapers were venturing online simply by replicating their print version for internet distribution—and often waiting until the publication of the print edition for internet release. It was obvious that more extensive disturbance of these industries was approaching, but competition from blogs or aggregators was still in an early stage. Little attention was paid to online-only journalism until the mid-2000s. *The Drudge Report* had played an important role in breaking news in the Lewinsky scandal in the late 1990s, but the scale of disruption really wasn't apparent until *The Huffington Post*'s launch in 2005.

The new media world wrought by internet distribution—at least in 2002—appeared beneficial for media consumers and destructive for legacy media businesses, which also explains the contradictory understandings of the state of television at the time. Perhaps it simply came down to this: night after night, new and unprecedented content undeniably appeared on the screen. No one was certain of how the internet would affect television—or had any idea that very soon many would carry pocket computers with HD screens that would let them watch "anything, anywhere." But it was impossible not to suspect that extraordinary changes were coming.

Following *La Femme Nikita* and HBO's end-of-century programming innovation, the next milestone in the story of how original scripted cable series transformed television comes from two series that debuted four months apart in 2002. *The Shield* and *Monk* provided the next volley in cable's onslaught toward legitimacy. The series were profoundly different, although both were arguably rooted in the well-tread path of the police procedural. Both shows would become iconic in the brand identities their channels would use to develop many subsequent series. Within just a few years these channels became as compelling sources for original scripted programming as any broadcaster or HBO. But in 2002, it seemed unlikely that these shows would become milestones in television history.

Independent from these technological changes, cable channels initiated a reimagining of television programming by continuing to pursue distinction as a strategy. The decade began with no meaningful original, scripted series production by advertiser-supported cable and ended with such abundance that the strategy of distinction began to fail because there was so much "distinctive" programming.

11 *The Shield*: Not Your Father's Cop Show

The Shield debuted in March 2002 on FX. Even before it appeared on screens, word of its edgy content circulated so widely that more people probably had heard that several advertisers pulled their commercials from the series than had actually seen early episodes.

The series was in some ways a marriage of *The Sopranos* and *NYPD Blue*. It pushed the police procedural into a new place by going beyond the tales of the Sisyphean nature of police work to place a criminal, as a protagonist, on the police force. *The Shield* tells the story of Vic Mackey, a career cop who polices by an "ends justify the means" strategy. Most of the series takes place in an inner-city Los Angeles precinct and follows Mackey and his "Strike Team," which was modeled after the "Rampart Division," whose widespread corruption scandalized the LAPD in the late 1990s. The Strike Team starts out as good cops, but the need to cover up misdeeds steadily escalates their corruption as viewers are drawn in by questions of when and if they will be exposed. The series also spends time showing Mackey as a family man and depicts him rationalizing his illegal acts as necessary to secure the needs of his family.

Like *The Sopranos, The Shield* (2002–2008) presented a character drawn carefully enough to make him sympathetic and able to compel audiences to follow his journey as he deliberated not between right and wrong, but between two wrong decisions.[1] FX brought as much of the HBO playbook of distinction to advertiser-supported cable as possible and offered a graphically violent series without a clear hero set in a world more morally ambiguous than *The Sopranos*.

Understanding how *The Shield* came to be requires looking at FX, the channel that developed it. FX is a much younger channel than most other top-tier cable channels. It launched in 1994, nearly twenty years after USA's

1977 debut. Because it began competing later, it was still struggling to gain carriage on many cable systems throughout the 1990s. Most notably, it was not available on Time Warner Cable in New York City until 1999.

FX drew little attention in its early years. Mostly, parent company News Corp. used it as a second market for the series its television studio produced for broadcast networks such as *The X-Files, Buffy the Vampire Slayer, Ally McBeal, NYPD Blue,* and *The Practice.* It also aired sports programming licensed by News Corp. that didn't have a place on Fox or other News Corp.–owned channels.

FX's motivations for developing original programming in the early 2000s were the same as USA's had been five years earlier. If FX wanted to improve its advertising rates and its subscriber fees, it needed cable subscribers to know it existed and care to have it. FX's strategy of acquiring series owned by its conglomerate proved a disappointing failure in the early 2000s. A profile of the channel published just before *The Shield*'s debut noted that corporate patience with the channel was waning and there was considerable concern about the substantial sums being spent on original programming.[2] FX needed a bold hit.[3]

The Shield was born less as a television show than as a creative exercise. Executives at Fox TV Studios—which mainly produced Fox's sitcoms—sent FX a script Shawn Ryan had submitted instead of the comedy script he was contracted to write. Ryan couldn't come up with a compelling comedy concept and wrote the first episode of *The Shield* to try to get past his writer's block, not imagining it would ever be produced. Although it was not what Fox TV was looking for, the studio executives were impressed with the script. The dark police drama was compelling enough that it became FX's first big drama gamble.[4]

The Shield was most certainly a swing for the fences. It more closely followed HBO's strategy of creating programming distinct from what could be found on broadcast than any other ad-supported cable channel had developed, and this quickly became FX's brand.[5]

The Shield established a strong audience from the start, but was not a breakout hit that drew widespread attention in its first season. Its debut set a record as the most watched original ad-supported cable series premiere with 3.1 million households, and it was the first time FX had ever ranked among the top ten cable programs of the week.[6] The controversy created by advertisers pulling out of early episodes gained the series notoriety, but the real attention came after an Emmy Award win for Michael Chiklis in the role of Vic Mackey after the first season and Golden Globe Award wins for Chiklis and for *The Shield* as Best Drama just as the second season began.

The awards attention was fortuitously timed as the second season debuted just as the first season was released on DVD. The release of a previous season of a series on DVD had just begun to be common, and it was particularly important in this case for enabling new audience members to catch up and join the show.[7]

FX followed *The Shield* with an uncommonly successful succession of series. Excepting *Lucky* (2003), a comedy about compulsive gamblers that emerged next and quickly faltered, FX's subsequent dramas *Nip/Tuck* (2003–2010) and *Rescue Me* (2004–2011) drew audiences, but what is more important is that they reinforced the edgy brand established by *The Shield* with really different shows.

Like HBO, FX followed its first success with a series that could not be considered topically related to its first. Rather, it established its brand through the tone of its series. Just as *The Shield* featured a protagonist who defied most everything American television had taught its audiences to expect from a protagonist, *Nip/Tuck* found few boundaries it wouldn't challenge. *Nip/Tuck* could be described as a medical series—with its over-the-top look at the world of two plastic surgeons—but it was outside of the well-established doctor/hospital subgenre. Stories of male characters' search for identity and happiness were at its core, as was the case in *The Shield* and several other cable series of this era. Instead of serving as the series' focus, these stories played out in family melodrama and weekly stories about the personal dissatisfactions of their patients.

Nearly a year to the day after *Nip/Tuck*'s debut, FX released *Rescue Me*, a post-9/11 series about NYC firefighters that was equal parts action, dark comedy, and personal drama. In many ways this series was closer to *The Shield* with its firehouse setting, themes of public servant work lives, and environment of male camaraderie. *Rescue Me* was also very different. Despite its self-destructive protagonist, open mourning of rescue workers who perished during the 9/11 attack, and scenes of heroic firefights, *Rescue Me* featured considerable humor. Like the other FX series, it was uncommon compared to the norms of broadcast television and unquestionably distinctive. Of the three series, *Nip/Tuck* attracted the largest audience with nearly three million in its third season and 3.4 million eighteen to forty-nine year olds for its fourth season premiere.

John Landgraf, who inherited the FX established by these three series in 2004, explained that "If FX hadn't had the great fortune to put three very commercially and creatively successful shows in a row on the air, then the whole of the business could have gone away. The fact that we achieved so much success on three successive shows caused this rush into the marketplace. Not only did Turner then rush into the market—following both us and USA—but then A&E and others tried to get in. And AMC will tell you directly that they just took a page from our playbook and chased after that very same model."[8] Landgraf's claim is bold, but not self-aggrandizing. It recalls precisely how tenuous the prospects of original, scripted cable series were at the time, even though these shows ultimately provided such a penetrating assault on broadcast dominance.[9]

Table 11.1
FX original series, 2000–2010

2000–2002	*Son of the Beach*
2002–2008	*The Shield*
2003	*Lucky*
2003–2010	*Nip/Tuck*
2004–2011	*Rescue Me*
2005	*Over There*
2005	*Starved*
2005–2006	*Thief*
2005–	*It's Always Sunny in Philadelphia*
2007–2008	*Dirt*
2007–2008	*The Riches*
2007–2010	*Damages*
2008	*Testees*
2008–2014	*Sons of Anarchy*
2009–2016	*The League*
2010	*Terriers*
2010–2015	*Louie*
2010	*Archer*
2010–2015	*Justified*

In the course of just a few years, FX went from being an obscure channel that even middling creative talent wouldn't deign to consider, to being *the* place to work. *The Shield* and its progeny were great shows, but are perhaps most important for extending the cultural and creative cachet of HBO to ad-supported cable. In 2001 cable channels were still considered the site of derivative, low-budget programming. Following series like *The Shield,* they quickly evolved into destinations of creative innovation.[10]

Television critic Alan Sepinwall rightly notes that "*The Shield* couldn't have existed without *The Sopranos,*" and argues that "in proving that you could do an HBO-style show away from HBO, *The Shield* was one of the most important series of the revolution."[11] *The Shield*'s importance derives from its success and from what it did to establish the FX brand. The series illustrated that much more creative artistry was possible on ad-supported television than broadcasters regularly offered.

The fact that this creative accomplishment was paired with significant commercial success for both its studio and FX led many to take note. As *Monk*'s simultaneous venture illustrates in the next chapter, following the model of HBO distinction was far from the only strategy available to ad-supported cable channels. The substantial difference between *Monk* and *The Shield*—despite emerging at the same time and enjoying similar long-term success—makes these shows and this moment so profound in the process of cable's transformation of television.

12 *Monk*: Just Distinct Enough

Over its eight-season run, *Monk* danced between broadcast and cable worlds in a way no other series has. *Monk* was first developed for ABC, but was ultimately created for USA. It is unquestionably a show a broadcast network wouldn't produce. Yet on two separate occasions, episodes of *Monk* aired during prime time on two different broadcast networks. These were most unusual situations, but they underline just how precisely *Monk* skirted the expectations of programming originating from a cable channel rather than a broadcast network.

In the early 2000s, USA's executives also recognized the value of distinction. USA had tried several original, scripted series since its 1997 success with *La Femme Nikita* (*USA High, Cover Me, The Huntress, The War Next Door, Manhattan, AZ*). None approached equivalent success. Of course, none was particularly distinctive, either.

In many ways, *Monk* was a throwback to 1970s series such as *Columbo* or British, novel-based, mystery series such as those of Sir Arthur Conan Doyle and Agatha Christie. The audience, along with the police force and Mr. Monk as a consultant, followed clues to solve the crime of the week, which Mr. Monk ultimately explained in the final act.

Monk's distinction—the feature that made it unlikely to air on a broadcast network—was its protagonist. Adrian Monk was once an excellent detective, but he had been dismissed from the police force after the grief caused by his wife's murder left him overcome by a range of obsessive-compulsive disorders. Monk didn't violate the sensibility that television protagonists must be heroic in the way Tony Soprano and *The Shield*'s Vic Mackey did. Presenting the character in a way that made him likeable, and made it clear that the series wasn't mocking mental disorder, was a tricky line to negotiate given the comedic overtones of the series.

The idea for *Monk* emerged out of an otherwise unsuccessful conversation between film producer David Hoberman and a writer for *Late Night with David Letterman* named Andy Breckman. During a lunch to discuss ideas for film projects, neither man offered a film concept that interested the other. The need to politely fill the rest of the lunch with conversation led to a casual discussion about television. Jack Nicholson's character in *As Good as it Gets* (1997) intrigued Hoberman, who speculated about whether a series involving a police officer with obsessive-compulsive disorders would make for good storytelling. Breckman, steeped as much in the mystery writing of Sir Arthur Conan Doyle and John Dixon Carr as in the comic arts, hit upon an idea that combined the two worlds immediately.

Hoberman had run Disney's Hollywood Pictures. When Disney decided to shut down the studio, it made several production commitments to Hoberman for future film and television work. Breckman recalls that they were consequently able to sell ABC on development of what would become *Monk* with a telephone call and a one-page description of the series. Breckman then endeavored to write the pilot script. All went well until they set about casting the leading role.

Adrian Monk proved to be an extraordinarily challenging role to bring to life. Although his compulsions had comic overtones, the series would be offensive if it seemed the show was making fun of his disorder. Also, many features of his personality made him grating and easy to dislike, not typical qualities of a leading character, especially on a broadcast network. The development team interviewed actor after actor and not one could find the delicate balance needed to humanely display the mix of grief, genius, and anxiety crucial to the character. Out of options, ABC gave up and shelved the project.

Although it is fairly common for a film to stop development and later reemerge and be produced—placed in "turnaround" in the industry vernacular—this was not commonly the case in television. Shelves were permanent resting spots for many television ideas that were simply before their time or that hit upon a development hurdle such as difficulty in casting a role. Breckman and Hoberman moved on and forgot about Mr. Monk. It was over a year later that Breckman received a call from Hoberman saying *Monk* might be coming back to life.

A junior executive at ABC named Jackie de Crinis—who Breckman doesn't recall ever meeting—had been involved with developing *Monk* at

ABC. She took the pilot script for *Monk* with her when she left ABC in 2000 for a job in scripted series development at USA. This USA was a world removed from the one that had persistently endeavored on the long effort to bring *La Femme Nikita* to life, although its final episodes had not yet aired.

De Crinis was hired at USA as a scripted development executive, but she quickly became uncertain of her role there. The executive team she reported to didn't view scripted originals as a necessity and instead prioritized the large audiences drawn by acquired series such as *JAG* and *Walker, Texas Ranger* that performed well for the channel at the time.[1]

There was good reason for this strategy. USA had stumbled with every original, scripted series it debuted since *La Femme Nikita*, and these were costly failures. No new series were actively being developed when de Crinis arrived, and she was not enthusiastic about any of the ideas available. A few months into the job, Stephen Chao, then president of programming and marketing, walked past de Crinis's office and asked her what she liked. De Crinis was surprised to find such a senior executive in her doorway and, unprepared for this question, handed Chao her copy of the *Monk* pilot script. Chao returned to his office and forty-five minutes later sent de Crinis an email calling it "delightful." De Crinis quizzed coworkers who had more experience working with Chao as to what this meant, but she soon had an answer. Chao returned to her office and told her "let's make this one."

And with that, *Monk* was reborn.[2] As chance would have it, an actor emerged who had been working on another series and had not read for the role when ABC was casting. Tony Shaloub had the skill to give Monk the soul the character required. He would receive eight Emmy nominations and win the award for best actor in a drama three times, as well as two Screen Actors Guild awards and a Golden Globe for his portrayal of this complicated character.

But finding a Monk wasn't the only staffing challenge. *Monk* was an unconventional series at a channel with little series experience. It was to be the first for this new team, and Breckman, who had never written or produced a scripted television series, was leading its creation.[3]

Breckman didn't set out to challenge the norms of television writing; he simply had never experienced those norms. And he assembled a writing staff as inexperienced with them as he was. Although a few had been staff writers on shows such as the *Late Show with David Letterman* or *Saturday Night Live*, none had ever written for a scripted series.

Monk was not like anything on television at the time, although many subsequent series have reproduced its formula nearly to the letter—a formula Breckman lifted from Doyle's Sherlock Holmes mysteries. In the early 2000s, police and detective shows were uniformly multicharacter ensembles. Their stories were driven by reviewing evidence and increasingly focused on crime scene forensics—this was the heyday of *Law & Order* and *CSI*.

In contrast, *Monk* focused on a single detective, used no high-tech science, and blended so many genres it is difficult to concisely describe the series.[4] Tonally, *Monk* was a comedy, but television comedies were still mostly confined to thirty-minute situation comedies. Its subject matter revolved around solving a mystery, though there was little precedent for comedic mystery solving. Moreover, *Monk* mixed established mystery subgenres. Unlike *Columbo*, an "open mystery" in which the audience sees the crime committed but is then compelled to continue watching by a desire to see how Columbo solves the case, audiences were compelled to watch episodes of *Monk* by the mystery of who perpetrated the crime, but also by how Monk would solve it. Finally, the mystery of Monk's disorder—which developed after his wife was killed—and that of her death added other layers to the storytelling. These features made *Monk* uncommon and suited USA's need to create a program of some distinction, while not drawing that differentiation nearly so starkly as FX.

By chance, de Crinis stumbled upon another show that "fell out" of development for another broadcaster as USA developed *Monk*. USA launched *The Dead Zone*—based on the Stephen King novel—a month before *Monk*. It initially drew 6.4 million viewers compared to *Monk*'s 4.8 million, but *Monk* continued to grow and was bolstered by its award nominations and wins.[5] Having the pair of series allowed USA to leap back into the cable competition.

Monk never drew the attention or accolades garnered by *The Sopranos*. Yet, it was just as massive a hit from a business perspective. *Monk* was a grand slam in terms of USA's hopes. Its numerous awards for acting and music lent legitimacy to the "elective" project of developing original, scripted cable series. *Monk* was important and especially successful not only for USA, but also for Universal Cable Productions, the studio that produced it and owned its rights (and notably a studio owned by the same company that owned USA).[6]

USA had no clear brand when *Monk* launched and was most identified with the American flag that was included with its logo—what USA's development executives jokingly referred to as "the postage stamp." USA followed *Monk* with several series that slightly adjusted the broadcast staple of the police procedural, but it distinguished its shows by centering them on characters that were unusual.[7] USA began using the slogan "Characters Welcome" in 2005 to indicate its identity, which was seen in series as disparate as *Monk, The Dead Zone, Psych,* WWE wrestling, and the various acquired series that also were a strong draw for the channel.

Monk inaugurated a subtle brand that also differentiated USA from other general-interest cable channels. USA hit its stride in 2007 and the "Characters Welcome" identity was clearest when *Monk* aired along with *Psych* (2006–2014) and *Burn Notice* (2007–2013). *Psych* adjusted the detective procedural slightly to place a detective's son others believed psychic—but really just uncommonly observant—at the center of weekly mysteries also told with a comedic tone. *Burn Notice* related the story of a former spy who used his skills to help those in need while trying to find out why he was burned. USA followed these series with a drama about the marshal service, *In Plain Sight* (2008–2013), a twist on the medical procedural, *Royal Pains* (2009–2016), and *White Collar* (2009–2014), in which a con man pairs with an FBI agent to help solve the type of white collar crimes he previously perpetrated.

Monk is important in understanding cable's transformation of television in a number of ways. Foremost, the show became an important beachhead behind which USA launched a steady stream of original series that established it as a site for original programming. When *Monk* debuted on USA in 2002, its producers could hardly have imagined that the show would span eight seasons and 125 episodes. They certainly didn't anticipate that its episodes would be sold to channels and networks in more than sixty-two countries, be released on DVD, and be sold in domestic syndication[a] to

a. *Syndication* is the term for selling a series in other markets after it airs on the outlet that licensed it. Domestic syndication historically was the selling of television shows to local stations. When cable emerged, it became an additional marketplace for selling series—especially dramas. Cable is considered a different window than domestic syndication, but functions similarly.

Table 12.1
USA original, scripted series, 2002–2010

2002–2006	*The Dead Zone*
2002–2009	*Monk*
2003	*Peacemakers*
2004	*Touching Evil*
2004–2007	*The 4400*
2005	*Kojak*
2006–2014	*Psych*
2007–2008	*The Starter Wife*
2007–2013	*Burn Notice*
2008–2012	*In Plain Sight*
2009–2016	*Royal Pains*
2009–2014	*White Collar*
2010–2014	*Covert Affairs*

multiple broadcast stations. In surprising ways, *Monk* followed many conventions of the business model standard for broadcast television: it served its original licensor—USA—well in its initial run, but it also returned significant revenue to its studio through sale of the series in several subsequent markets. These sales kept *Monk* in circulation long after production ended and allowed many more viewers to see it.[9]

Monk attracted viewers and critics who came to regard it just as they would a broadcast series. It provided the cornerstone of a viable and valuable brand strategy that demarcated USA as offering programs distinct from broadcast networks but still of general interest and broadly targeted. Moreover, it established USA as a place to look for award-winning, original series.

The aspect of *Monk*'s production that remains unique is that the series aired on both ABC and NBC during its original production. The ABC airing came about because of its unusual development history and a rarely invoked aspect of the boilerplate agreement that allowed USA to redevelop *Monk* after ABC passed on it. Buried deep in the legal language was a provision that gave ABC the right to air episodes after their original appearance on USA. ABC's schedule was in such dire straits in the early 2000s that the network invoked this provision to see if "rerun" episodes of *Monk* could improve its fortunes.

Then—six years later—at the conclusion of the 2007–2008 Writers Guild strike, NBC—then the owner of USA and the studio that produced *Monk*—scheduled previously aired episodes of *Monk* to provide some scripted content while production resumed on series that had been on hiatus during the strike.[8]

Monk, then, was an enormous success for both USA and its studio. As such, it approximates the norms characteristic of conventional broadcast practices more than many other milestone series in this history. The success *Monk* ultimately achieved can be attributed to all the things that made it seem most risky at the time—the inexperience of Breckman and his writing staff, its focus on a single character, an unusual protagonist, its blending of genres, and its defiance of typical television mystery structure.[10]

Like most of the other series considered here, no one would have predicted *Monk*'s status as a milestone in television history. Almost everything about its development was uncommon, and a few things were unique. But comparing *Monk* with *The Shield* reveals how their uncommonness may have enabled these shows to become path-charting rather than notorious failed experiments. Creating scripted series for cable was an uncertain enterprise. That uncertainty allowed, and even required, unconventional practices. Many of these unconventional practices led to the creation of series that were unimaginable when television was defined by broadcasting, but they immediately set a new bar that led others to pursue yet further innovation.

Each in its own way, *Monk* and *The Shield* established that advertiser-supported cable series could match and exceed broadcast excellence, although they did not quite reach the bar set by subscriber-funded HBO. *Mad Men* would challenge this standard later in the decade, but the awards earned by *Monk* and *The Shield* began the assault. *Monk*'s finale episode remained the highest-rated episode of a scripted cable series until *The Walking Dead*.[11]

The success of *The Shield* and *Monk*—series that cemented effective strategies for their channels for the next decade—indicated the arrival of a specific but brief phase in the transition of television programming norms. At the end of their runs, *The Shield* and *Monk* departed from a television environment very different than the one they in which they debuted. Many commentators may have suggested television was dying during the period from 2002 to 2010, but it was simply that cable channels were upending broadcast television's dominance.

Audiences became accustomed to looking beyond broadcast networks for new and interesting series surprisingly quickly after the launch of *The Shield* and *Monk*. Within a decade of the debut of *La Femme Nikita* and *OZ*, original, scripted series completely erased cable's secondary status. In the early 2000s, only USA and FX succeeded in developing distinctive original scripted series. Notably, others also tried.

13 Cable's Rising Tide Doesn't Lift All Channels

More narrowly targeted channels such as Lifetime ("Television for Women") and Sci Fi (defined by genre and rebranded as Syfy, with the tagline "Imagine Greater," in 2009) developed series successful with particular audiences in these early years of original cable series, but they didn't have shows that gained broader recognition. A&E attempted prestige drama with the Sidney Lumet–created, legal drama *100 Centre Street* (2001–2002) and an adaptation of the *Nero Wolf* (2001–2002) period detective stories, while Bravo aired the newsroom drama *Breaking News* (2002). In comparison with *Monk* and *The Shield,* these quickly canceled series could be described as both more and less conventional—suggesting the delicate balance of difference and similarity needed for commercial success.

The most-watched cable channel, TNT, struggled to develop original series. TNT rebranded in 2001 to specify the more particular brand of "TNT: We Know Drama." TNT developed a range of series such as the Wall Street–set *Bull* (2000) and the fantasy/adventure *Witchblade* (2001–2002). But the channel's original series drew little notice until 2005 with series that subtly varied from the police procedural common on broadcast channels. TNT launched its most successful bid to compete in the original scripted series race with *The Closer* (2005–2012), followed by *Saving Grace* (2007–2010), *Southland* (2010–2013), and *Rizzoli and Isles* (2010–2016). These shows were largely traditional police procedurals but, like *Monk,* they featured protagonists who deviated from the conventions of broadcast series. TNT's success with such shows indicated an evolution of U.S. television's competitive field.

Though TNT at last joined general-interest competitors FX and USA in scheduling successful scripted originals, much of the rest of the ad-supported cable landscape continued to struggle in this pursuit. Ironically,

Table 13.1
Early 2000s live action original scripted series on ad-supported channels other than USA and FX

2000–2001	*18 Wheels of Justice*	TNN
2000–2002	*The Invisible Man*	Sci Fi
2000–2001	*Strip Mall*	Comedy Central
2000–2006	*Strong Medicine*	Lifetime
2000	*Bull*	TNT
2001	*Black Scorpion*	Sci Fi
2001	*That's My Bush!*	Comedy Central
2001–2002	*The Chronicle*	Sci Fi
2001–2002	*100 Centre Street*	A&E
2001–2002	*Nero Wolfe*	A&E
2001–2002	*Witchblade*	TNT
2001–2005	*Andromeda*	Sci Fi
2001–2004	*The Division*	Lifetime
2002–2003	*For the People*	Lifetime
2002	*Breaking News*	Bravo
2003	*Tremors: The Series*	Sci Fi
2003–2009	*Reno 911!*	Comedy Central
2003	*Playmakers*	ESPN
2003–2005	*Wild Card*	Lifetime
2003–2006	*Missing*	Lifetime
2003–2010	*Battlestar Galactica*	Sci Fi
2004	*Good Girls Don't*	Oxygen
2004	*Significant Others*	Bravo
2004–2009	*Stargate: Atlantis*	Sci Fi

channels such as Lifetime that had shifted to targeting specific audiences and seeking program distinction in the mid-1990s lost ground throughout the 2000s.

Lifetime had emerged from nowhere in the early 2000s to claim the title of most-watched cable channel from January 2001 to March 2003. Lifetime's rise resulted from a new president who quintupled the channel's promotion budget and expanded programming funding by 20 percent. It also didn't hurt that Lifetime's "niche" potentially encompassed half the viewing audience.

Within two years the channel jumped from sixth to first in prime-time viewership.[1] By 2003, it had dropped back to sixth, less a result of its programming (although it did have difficulty launching new originals after 2001), but because its general-interest competitors developed original series and more clearly defined brand identities.[2]

The brief tenure of Lifetime's top ranking was also due to broadcast networks pivoting in their programming strategy. Viewers prefer programs targeted to their specific interests and perspectives. In the days of only broadcast networks, few programs were targeted with much specificity. The competitive environment was made up of just a handful of channels that were all trying to appeal to as many people as possible. Cable channels could succeed with smaller audiences because they were funded by subscriber fees as well as advertising dollars. By providing targeted programming, cable (and the upstart broadcast networks) changed the competitive strategies for all of television.

Broadcasters also began narrowing their targets. With hits at the dawn of the twenty-first century such as *Judging Amy, Crossing Jordan, Grey's Anatomy,* and *Desperate Housewives,* broadcasters were emboldened to explicitly target female audiences to an unprecedented degree. Developing a show unlikely to appeal to men was no longer a big risk for a broadcaster. Men were already watching something else. The broadcasters might as well deliberately target women.

Unlike general entertainment channels like FX and USA that needed creatively distinctive programming to differentiate themselves from broadcast networks, Lifetime relied on its female focus as its distinction. The channel offered programs much less creatively inspired. Many followed a formula that involved airing low-budget versions of broadcast fare with mostly female casts or settings. After seeing Lifetime's success, other general-interest channels such as TBS and TNT adopted more specific brands (comedy and drama respectively), although USA remained broadly focused. At first these brands merely guided acquisitions but, in time, they led to the original scripted series these channels developed.

Lifetime wasn't the only demographically targeted channel that struggled in the 2000s. After slowly evolving from The Nashville Network (1983–2000) to The National Network (2000–2003), owner Viacom announced in 2003 that TNN would become Spike, "television's first entertainment network for men." Spike's quest to become television for men reveals a lot about the challenges and competitive dynamics of cable throughout the first decade of the twenty-first century.

Despite announcing itself as a men's television network, Spike struggled in this pursuit because little of its programming clearly communicated that identity. Like many channels, most of the hours of its schedule were filled with shows originally produced for broadcast networks. Such broadcast-developed shows targeted both men and women. An average day just after its launch in September 2003 was filled with familiar acquired programs: *Baywatch, Miami Vice, The A-Team, Real TV Renewal, Seven Days, Star Trek: The Next Generation, Highlander,* and *Blind Date*. At best, Spike aggregated a collection of programs that skewed to male audiences on a single channel.

The biggest challenge to its identity as a channel for men was the purchase of *CSI* episodes, a deal made before the rebranding. The old *CSI* episodes drew a significant number of new viewers to the channel. Ironically, though, it drew a lot of women. So much so that Spike became the sixth-most-watched cable channel among women.[3]

Spike's first original programs were also developed before the announced rebrand, but they were suggestive of the channel's new identity. The animated series *Gary the Rat,* about a lawyer transformed into a rat, *Stripperella,* a stripper voiced by Pamela Anderson Lee who moonlights as a super hero, and *Ren & Stimpy: Adult Party Cartoon* were commissioned for the pop-culture brand of The National Network, even if some of this programming (particularly *Stripperella*) clearly exhibited Spike's decision to be a channel for men with a laddish, *Maxim*-esque sensibility.[4]

Spike's struggles were particularly evident in comparison with the brand FX established in these same years. While Spike made splashy announcements claiming itself as television for men, FX actually developed this reputation with its succession of original series including *The Shield, Nip/Tuck,* and *Rescue Me*. Spike even purchased episodes of *The Shield* after they aired on FX because of their fit with the channel's intended brand.

Spike struggled to gain traction and evolved its television for men brand through a number of subtle shifts, ultimately settling on the tagline "Get More Action." Despite allowances for construing this as a double entendre resonant with Spike's earlier laddishness, the network's interpretation seemed decidedly literal.[5] In early 2017, owner Viacom announced that Spike would be rebranded as a general entertainment channel called the Paramount Network.

This accounting of Spike's conflicting brand and programming i meant as critique as much as to highlight a different story about a ca channel during these years in which original scripted series drew so muc attention. Far more channels struggled with the new competitive environment than succeeded, even those that developed original scripted series such as Lifetime and Spike. The need to fill a twenty-four-hour schedule with on-brand programs challenged cable channels. Cable revenues were not significant enough to fill a daily schedule with original programming. This meant that much of a channel's schedule remained full of old broadcast programs even if it was able to develop one or two original series. These old broadcast programs consistently impeded efforts to establish a clear brand.

This phase of television's evolution would have been quite different had streaming and video on demand been prevalent at this point. Branding cable channels would have been far easier if they could have simply existed as on-demand libraries of original programming instead of channels requiring a full daily schedule even though the business of television would have seemed to be in considerable crisis without cable channels spending millions, even billions, on buying former broadcast series. Cable channels' "scheduled" delivery led to a highly inefficient set of business practices that would soon be the prey of internet distribution.

14 Cable Gets Real

Of course, scripted originals were not the only programming strategy that a cable channel could use to draw audiences and articulate a brand identity. An abundance of original scripted cable series aired by 2010, but only a small percentage of cable channels—fewer than two dozen of more than five hundred national cable channels—had endeavored to develop these programs. Most cable channels limited their original programming to what were termed *alternative* series in the industry, better known to audiences as *reality television*.

Alternative designates a broad category that encompasses a wide array of forms. It is perhaps easiest to distinguish alternative series from what they are not—they are not performed by actors following a script, and they are not films.[1] Original alternative series provided a tool for cable channels to communicate their brand and to provide some original content amid schedules typically packed with programs previously aired by broadcasters. Channels used them in the same ways and for the same reasons some turned to scripted originals. The key advantages of alternative series resulted from the fact that they were relatively cheap and quick to produce. Several channels found lightning in a bottle with an alternative concept. The ultimate value of those successes for their channels varied considerably.

Even the biggest alternative hits—shows such as *Deadliest Catch, Jersey Shore,* and *Duck Dynasty*—enjoyed great notoriety for only a brief period. Few successful alternative series sustained their popularity for multiple years. Alternative series were consequently of limited use for cable channels hoping to develop programming that would help increase the fees cable service providers paid and to ensure continued carriage. These agreements were renegotiated every five to eight years, so a short-lived hit meant little in these negotiations. Certainly a popular ongoing alternative series

(*MythBusters, The Daily Show*) could draw attention to a channel that might make viewers notice if it disappeared from a cable provider's options. A track record of consistent alternative development, such as that achieved by MTV and Bravo, also could be valuable.

The alternative series that became hits were often somewhat different from the channel brand. This limited their value in attracting viewers to other shows on the channel. How many viewers remember that *Ice Road Truckers* was a hit for The History Channel or that *Storage Wars* was on A&E? Successful series quickly became bigger than their channel and developed significant popular culture awareness that was divorced from it. A big alternative hit could also lead a channel to lose its way and adjust its brand to follow a successful program. Once the series passed from fashion, the channel was left with a diluted brand.

A popular alternative series could prove valuable in terms of advertising revenue, although here too gains were limited. The outrageousness that often drew audiences to alternative series also made advertisers nervous.[2]

Finally, alternative series were much cheaper to produce, but they had little value after their original airing.[3] The alternative shows that succeeded on cable were less likely to be sold as a format that could be remade in another country than the type of alternative series common on broadcast networks (*American Idol, The Bachelor, Survivor*), although there were exceptions (*Real Housewives*). Cable channels that owned channels in other countries often produced a local version for channels outside the United States. For example, MTV produced Latin American, Spanish, and British versions of *Jersey Shore*, although it also marketed the U.S. version of the series on several of its international channels.[4]

To be sure, alternative series represented a much larger percentage of original cable series production than scripted series, but they are considered only slightly here because of their limited impact on the broader transformation of the business of television. Broadcasters had their own alternative series revolution—focused mostly on broadly targeted competitions such as *Who Wants to Be a Millionaire?, Survivor, American Idol*, and *Dancing with the Stars*—that began a few years before cable hit its stride with its alternative series. The alternative format that was most successful on cable—the docusoap—remained a mainstay of cable with few examples and no breakout successes on broadcast networks. Cable channels' alternative series proved massively successful—many of the biggest hits doubled and even tripled the audiences of successful scripted cable series in this period—but the successes had constrained impact on the overall business of the channels and for U.S. television in general.

Within the extensive diversity of alternative series, though, the attention to sensational shows makes it easy to overlook subtler strategies. Daily offerings such as ESPN's *SportsCenter* and Comedy Central's *The Daily Show* were often left out of assessments of alternative programming. There is a fine and uncertain distinction between alternative series and talk shows, and talk shows had long been the most common original programming on cable channels. The regularity and frequency of talk shows were crucial to their channels. Such series precisely reproduced their channels' brands and cultivated regular viewing that was of critical value when the channels' owners renegotiated carriage deals and fees with service providers.

Renovation/DIY series also quietly produced steady value for channels such as HGTV and the Food Network. Unlike docusoaps and ongoing competitions, shows such as *House Hunters* and *Iron Chef* did repeat well. They may not have inspired appointment viewing or drawn comparable pop culture buzz, but they attracted steady audiences over time and could be produced on a modest budget. This category of alternative series has a longer history and is characteristic of some of the earliest original cable content.

The phenomenon of original alternative series that drew uncommonly large audiences to cable channels and earned significant cultural attention began with MTV's *Real World*. *Real World* launched in 1992 and reached its peak in viewership in 2004 when four million viewers watched its premiere. *Real World* was an outlier for a decade. It wasn't until MTV's *The Osbournes* debuted in 2002 and also gathered four million viewers that docusoaps began to develop as a program form other cable channels replicated.[5]

Noteworthy alternative series development extends beyond MTV in 2003. This was a watershed year for original cable alternative series by many measures. Bravo launched *Queer Eye for the Straight Guy*, which doubled its ratings to 2.2 million over an eleven-week period and drew extensive attention in an era in which openly gay characters and personalities remained uncommon on television. Discovery first aired *Mythbusters* and *American Chopper* in 2003, extending the array of alternative formats, while MTV continued its alternative development with *Punk'd* and *Newlyweds: Nick and Jessica*. Neither was as successful as *Real World* and *The Osbournes* had been in expanding the MTV audience, but with 2.7 and 2.4 million viewers, respectively, both provided a solid audience for the niche brand. The trend continued in 2004 with the debut of *Laguna Beach* on MTV and *Project Runway* on Bravo. *Deadliest Catch,* which typically drew three million viewers and spawned several other "men doing challenging jobs" series, premiered in 2005 on Discovery.

The primary alternative formats were established by this point: docusoaps—*The Osbournes* and *Laguna Beach*; makeover shows—*Queer Eye for the Straight Guy* and *What Not to Wear*; and cable's version of the gamedoc—*Project Runway*. Several of these series—as well as A&E's *Growing Up Gotti*—drew over three million viewers, while many other shows drew audiences in the one to three million range, which could be a significant audience for more narrowly targeted cable channels (*Pimp My Ride*, MTV; *College Hill*, BET; *Dog the Bounty Hunter*, A&E).

Several cable alternative series drew audiences comparable to cable scripted series by 2007, but the lightning-in-a-bottle shows also began to emerge. In the latter category, VH1's *Flavor of Love* finished its second season in 2007 with 7.5 million viewers, which made it the year's most-watched nonsports, advertiser-supported cable show. Also that year, *A Shot at Love with Tila Tequila* (MTV) attracted 6.1 million and VH1's *Rock of Love* and *I Love New York* drew over five million for their finales. Several other shows that would later become blockbuster hits launched in 2007: *Jon and Kate Plus 8* (TLC) reached 9.8 million in its fifth season, and *Keeping Up with the Kardashians* debuted on E!, although it didn't achieve its highest rated episode, seen by 4.8 million viewers, until 2010. Likewise, *Jersey Shore* premiered in 2009 and drew nearly five million for its first season finale, but grew to a peak of 8.7 million in its final season. History's *Pawn Stars* debuted in 2009 and reached a height of 7.6 million in 2011. The record-setting *Duck Dynasty* (A&E) launched to a mere 2.5 million in 2012 before amassing 11.7 million viewers for its fourth-season debut.

A couple of important stories are embedded in this brief history. First, some channels developed a steady stream of original alternative series that created a consistent audience. A few channels were able to quickly replace one-time hits that had passed their prime and sustain the audience with other series. This is what makes alternative series a valuable strategy.

MTV arguably best exemplified such a steady approach. The channel introduced regular innovation—initiating many formats or quickly

Table 14.1
Major cable alternative series

1992–	*Real World*	MTV
1994–	*Road Rules*	MTV
1997–2004	*The Crocodile Hunter*	Animal Planet
2000–2008	*Trading Spaces*	TLC
2002–2005	*The Osbournes*	MTV
2003–2007	*Nashville Star*	USA
2003–2007	*Punk'd*	MTV
2003–2007	*Queer Eye for the Straight Guy*	Bravo
2003–2005	*Newlyweds: Nick and Jessica*	MTV
2003–2013	*What Not to Wear*	TLC
2003–	*American Chopper*	Discovery
2003–	*Mythbusters*	Discovery
2003–2006	*The Surreal Life*	VH1
2004–	*Project Runway*	Bravo, Lifetime
2004–2006	*Laguna Beach*	MTV
2004–2005	*Growing Up Gotti*	A&E
2005–	*Deadliest Catch*	Discovery
2006–2008	*Flavor of Love*	VH1
2006–2010	*The Hills*	MTV
2006–	*Top Chef*	Bravo
2006–2016	*Real Housewives of Orange County*	Bravo
2007–2011	*Jon and Kate Plus 8*	TLC
2007–2009	*A Shot at Love with Tia Tequila*	MTV
2007–2009	*Rock of Love*	VH1
2007–2008	*I Love New York*	VH1
2007–	*Keeping Up with the Kardashians*	E!
2007–2016	*Ice Road Truckers*	History
2009–2012	*Jersey Shore*	MTV
2009–	*Pawn Stars*	History
2009–2014	*16 and Pregnant/Teen Mom*	MTV
2009–2016	*Toddlers & Tiaras*	TLC
2009–	*RuPaul's Drag Race*	Logo/VH1
2010–2013	*Basketball Wives*	VH1
2012–	*Duck Dynasty*	A&E

identifying how to adapt a format for its audience: the celebrity docusoap (*The Osbournes, Newlyweds*), the "regular" person docusoap (*Sorority Life, Laguna Beach, Jersey Shore*), and the dating competition spectacle (*Shot at Love*). MTV's youthful target and the way the channel capitalizes on audiences in a certain phase of life made this strategy more viable.

In contrast to MTV's consistent success, many big alternative series were briefly popular and then quickly faded away. Such abrupt successes could yield two outcomes for a channel. Some were able to exploit the attention and use it to develop and promote a clear channel identity. In 2003, *Queer Eye for the Straight Guy* was a noisy, attention-grabbing show that put Bravo on the map. Bravo followed *Queer Eye* with *Project Runway* and various celebrity-driven shows (*Being Bobby Brown*; *Kathy Griffin: My Life on the D-List*). In 2006, the channel developed *Top Chef* and *Real Housewives of Orange County*, which it extended into many different *Real Housewives* versions. The sustained development enabled Bravo to build an identity that stood out amid a crowd of cable channels.

Other channels struggled to use a breakout hit as the basis for a sustainable brand, however, and they fell back to obscurity after a peak show. *Jon and Kate Plus 8* drew widespread attention to TLC, a channel previously known for programs such as *Trading Spaces*, *A Baby Story*, and *Junkyard Wars*. The channel followed the show that became *Kate Plus 8* with other series focused on the inherent drama of family life that garnered initial attention, such as *Little People, Big World* and *17 Kids and Counting*, but it also launched *Cake Boss* as well as *Toddlers and Tiaras* and its spinoff, *Here Comes Honey Boo Boo*. In the process, it failed to establish a clear or consistent brand and lost much of the audience drawn by its broader hits.

Most channels focused their development on either alternative or scripted series. With the exception of A&E, no channel emphasizing docusoaps broke into the most-viewed channels—although many targeted such narrow audiences that this would never be possible. For example, MTV is not a top-ten channel, but it used alternative series successfully.

Alternative series took an array of forms and were used in many different ways. They could be just as distinctive as scripted series. A low-cost series that did break out brought attention to its channel as the field of competitors expanded. Series that could be repeated were especially valuable. The frequently outrageous docusoaps had the capacity to get people talking, pushing viewers who wanted to be in the know not only to watch, but to watch live. Live viewing meant more advertising views, which also increased the value of these series.[6]

Although the value returned by alternative series varied, many of these series drew significant cultural attention. Like the original scripted series, they played an important role in leading Americans to rethink their expectations of cable. If any lingering uncertainty about "cable" as a form of television remained by the summer of 2007, it was soon eliminated.

15 *Mad Men* Brings AMC Prestige but Loses Money

By 2007—just a decade after *La Femme Nikita* and *OZ* uncertainly debuted—expectations of cable had changed tremendously. The different perceptions of broadcast and cable that were profoundly acute in the mid-1990s had all but faded. Original cable series had grown common, and viewers were just as likely to expect that a new series might be found on a cable channel as a broadcast network.

There had been many scripted, original cable series by this point, although far more had failed than succeeded. Broadcasters still looked at the ratings of cable series with amusement. It was with barely contained annoyance that broadcast network programmers reminded journalists—who wrote disproportionately about series watched by few—that none of the programs they regularly fawned over ranked among even the top one hundred most viewed series in any given week.

It remained the case that any cable series drew a small fraction of the audience of even a mid-performing broadcast show. TBS's *House of Payne* debuted in 2007 as the most watched cable comedy ever, with 5.82 million viewers. Compared with broadcast series that year, that audience ranked 116. Journalists noted high ratings for several cable series in 2007—*The Closer* drew 7.3 million on TNT, *The Starter Wife* debuted to 5.4 million, and *Burn Notice* reached 3.6 million. Compared with broadcast series, 7.3 million viewers ranked *The Closer* seventy-ninth, and its audience size was exceptional for a cable series.[1] Cable original series ranked somewhat better if evaluated based on how many eighteen- to forty-nine-year-old viewers they drew rather total viewers, but the top fifty broadcast shows still drew more of this advertiser-coveted demographic.

There were now so many cable channels and cable shows that no one show drew a significant number of viewers, but the abundance of small audiences added up. The effect of the growing cable competition was curious. With the exception of sports—which drew the largest audiences to both broadcast and cable—broadcast network shows remained by far the most-watched programs. The array of top cable channels each siphoned a couple of million viewers away from broadcasters, but no cable show before 2010 caught the fancy of many more. This allowed the broadcast networks to remain something like a three-hundred-pound gorilla (reduced from their eight-hundred-pound glory days) being attacked by a disorganized pack of dogs. This strange disjuncture in which cable shows drew considerable attention and awards, yet were viewed by comparatively small audiences, fed the confusion about what was happening to television in the early years of the 2000s.

By mid-decade, competition among cable channels had grown more complicated. HBO began to stumble with a succession of series regarded as balance sheet failures (*Deadwood, Carnivale, Rome, John from Cincinnati*). Besides *The Shield* and *Monk*, no ad-supported cable series garnered major awards, and only Sci Fi's *Battlestar Galactica* drew significant critical attention. Cable channels had proven they could make good series, but there was some question about whether HBO's early success had been beginner's luck and whether ad-supported cable could develop a creative and commercial game changer.

AMC answered those questions in July 2007.

AMC, or American Movie Classics, launched in 1984 as a subscriber-funded channel that aired unedited, uncut, pre-1950s films. It dropped the subscriber fee three years later and then added advertiser support in 1997. At first it aired commercials only between films, but then added commercial breaks within films beginning in 2001. The channel relied considerably on nostalgia for its brand of "TV for Movie People."

AMC ranked twenty-first among cable channels in prime-time viewership when *Mad Men* debuted.[2] It had not attempted original scripted series since *Remember WENN* and *The Lot* a decade earlier. Uncommon viewership and critical success achieved by its 2006 original miniseries *Broken Trail*, which attracted 9.7 million viewers—nearly three times AMC's previous record—motivated the channel to develop programming that would attract more attention.[3] Rob Sorcher, AMC's head of programming and production, recalled being directed by the channel's CEO that "we need a *Sopranos*."[4]

The creation of *Mad Men* is as unusual as the other series considered so far. Creator Matt Weiner wrote the pilot script in an effort to prove his dramatic credentials in 1999. He had established a career as a comedy writer, but had difficulty even being considered for jobs on dramas. *Mad Men* began as a way to get Weiner off the unremarkable Ted Danson comedy *Becker*. Weiner's agent sent the script to David Chase in 2002, when he was in the midst of *The Sopranos*. It earned him an interview and then a job.

Weiner spent three successful years writing on *The Sopranos*. Meanwhile, the *Mad Men* script floated around and was considered, but dismissed by, at least, HBO and FX.[5] Weiner's manager gave the script to AMC after the channel put itself on the map with *Broken Trail*. Christina Wayne, AMC's senior vice-president of scripted programming, found it a compelling fit with the channel's classic films and reminiscent of the novel *Revolutionary Road*, which she had tried to develop years earlier as a screenwriter.[6] *Mad Men* didn't connect to the channel's films as clearly as had *Broken Trail*, but Wayne's superiors had enough confidence to push ahead with it.

Mad Men was set in the 1960s, and it told the complicated story of its lead character's struggle to adapt through this turbulent decade. Like FX dramas such as *The Shield*, *Nip/Tuck*, and *Rescue Me*, *Mad Men*'s protagonist, Don Draper, wasn't clearly a good guy. Don was an "ad man"—at his best making up a fantasy world for clients' products. *Mad Men* also had an uncommonly large cast for a cable series. It ultimately told detailed stories about several other characters, although it was primarily presented as Don's story.

AMC's development team knew they needed a signature show and something different than what could be found on other cable channels.[7] But there were many challenges to producing a series at a cable channel ranked twenty-first. AMC looked for a studio to produce the series but struggled to find a partner; AMC was not part of a conglomerate that owned a studio.[8] Sorcher recalled that he couldn't even get major studios to read the script.

Beyond the fact that AMC wasn't perceived as a significant player in cable, *Mad Men*'s commercial prospects were doubtful. The script may have exhibited all the brilliance it was later credited with, but *Mad Men* was a very American story, a period-specific story, and even an industry-specific story. Worse yet, the series was heavily serialized—more like a soap opera than a story contained in each episode. This made the series less attractive to channels that would buy it after it had aired on AMC because they often air episodes out of order and viewers are unlikely to watch regularly enough to follow the story. All of these components—as well as the fact that it was originating from an unknown U.S. channel—made its prospects for big sales in international, cable, or domestic syndication slim. There just wasn't a clear financial upside to interest the big studios that mainly produced broadcast series—a situation reminiscent of the challenge *La Femme Nikita* faced.[9]

AMC struggled both because it was unknown as a programming outlet and because *Mad Men* was an unconventional series. Part of the difficulty in selling a studio on taking such a significant gamble came from the fact that much of the brilliance of the series wasn't on the page, but in the visuals. AMC first approached the smaller studio Lionsgate TV to produce the series. The television unit of what was mostly a film company, Lionsgate didn't have as extensive an infrastructure as the major studios. Where a series such as *Mad Men* that might ultimately return only $50,000 per episode was too insignificant to the major studios, such returns were worthwhile to a studio like Lionsgate.

But top executives at Lionsgate initially viewed *Mad Men* as too risky. Hoping a completed episode might convince studios of the series' potential, AMC decided to finance the production of the first episode itself. Armed with the pilot episode, advocates for the project at Lionsgate were better able to argue the potential of the still-risky series to senior decision makers. Soon after, the parties reached an agreement for Lionsgate to produce it.[10]

A key aspect to understanding AMC's persistence with the series is that it explicitly was aiming for an exceptional show that would bring prestige. Having a ratings hit was a secondary consideration.[11] *Mad Men* wasn't a clear hit out of the gate; in truth, the series was never a hit by conventional measures. The series debuted to 1.6 million viewers but had fallen to 841,000 by its third episode. Nevertheless, AMC decided to renew it for a second season, which was a prescient move, as the series became an enormous critical and award success.[12] It won the prestigious Peabody Award in its first season and then went on to win Emmy Awards for best dramatic series in its first four seasons, a BAFTA for best international show, and several other accolades. But the ratings for the series were never extraordinary. The series averaged about two million viewers over its run—at the low end even for cable series. *Mad Men* may not have been widely watched, but it put to rest questions about whether cable—especially ad-supported cable—was capable of developing sophisticated television.

Someone living in the United States from 2007 to 2015 could be forgiven for believing *Mad Men* was an enormous cultural phenomenon. As had been the case with *The Sopranos* before it, the series caught the fancy of journalists and entertainment pundits; the number of articles published about *Mad Men* bore little relation to the relative size of its audience.

Many of the journalists writing about the series questioned the "profitability" of *Mad Men*.[13] The uncertainty resulted from its small audience, which meant limited advertising revenue. *Mad Men* likely lost money for AMC, yet, paradoxically, it was simultaneously very profitable. *Mad Men*'s value to AMC derived less from advertising than from helping it become a top-of-mind cable channel. By the end of *Mad Men*'s run, AMC improved its ranking to be the tenth most watched cable channel.[14]

Mad Men cost AMC about $26 million a year. Industry analysts estimated the channel received about $22,000 dollars per thirty-second commercial for the first airing of *Mad Men* and $12,000 for rerun airings.[15] At twenty ads an hour, and two rerun airings, *Mad Men* might earn AMC about $12 million in advertising per season—less than half its initial $2 million per episode cost. And that doesn't include the channel's promotion costs.

But ads supply only half of a cable channel's revenue. The motivation for cable channels to develop original scripted series—dating back to the changed competitive environment that emerged in 1996—was to increase the fees cable service providers paid to cable channels. Between *Mad Men*'s 2007 debut and 2010, the fees cable providers paid for AMC increased by two cents per subscriber.[16] That may sound insignificant, but that two cents multiplied by the 95 million homes that received AMC—regardless of whether they watched it—multiplied by twelve months, equaled $22.8 million a year.

The increase in the subscriber fee explains why cable channels gamble on original series and how a niche hit such as *Mad Men* could prove lucrative. In 2011, after *The Walking Dead* premiered on AMC and became a blockbuster hit, the subscriber fee grew an additional fifteen cents, which translated into an additional $1.7 billion of subscriber revenue.[17] This was especially beneficial in the valuation of AMC Networks as it went public in 2011.

Mad Men provides a useful lens through which to view the complicated economics of cable series, but its bigger legacy in the story told here is its creative accomplishment. *Mad Men* provides a milestone in this history as the first series to bring the acclaim of HBO series to advertiser-supported cable. Several subsequent series would quickly make audiences forget any perceived inferiority of cable originals. On the business side, *Mad Men* helped catapult AMC up the ranking of subscriber fees yielding millions in revenue.[18]

Extraordinarily, *Mad Men* accomplished everything AMC could have hoped. It helped AMC quickly become a recognizable cable channel that was then able to develop subsequent series that yielded greater success than failure: *Breaking Bad*, 2008–2013; *Rubicon*, 2010; *The Walking Dead*, 2010–; *Hell on Wheels*, 2011–2016; *The Killing*, 2011–2013; *Low Winter Sun*, 2013; *Halt and Catch Fire*, 2014–2017; *TURN*, 2014–; *Better Call Saul*, 2015–; *Fear the Walking Dead*, 2015–; *Into the Badlands*, 2015–; *Feed the Beast*, 2016; and *Preacher*, 2016–. By 2010, AMC had a show that attracted the largest audience on cable and one that regularly topped broadcast shows in the eighteen- to forty-nine-year-old demographic most valued by advertisers. That show—*The Walking Dead*—provides the next chapter of the story of cable transformation.

16 *The Walking Dead* Redefines Cable Success and Strategy

By 2010, the second part of this story—of how the internet revolutionized television—began to take shape. Before moving to those developments, however, a few more pieces of the story of cable's transformation of television remain to be told. Independent of the arrival of internet-distributed television, the competitive environment began shifting again. By this point there were so many original scripted series, many of which were of considerable distinction and accomplishment, that it became difficult to single out particularly noteworthy series.

Despite the abundance of distinctive shows, *The Walking Dead* stands out as the next installment in the story of how cable transformed television. Its predecessors proved that series produced for cable—even ad-supported cable—could achieve television's highest levels of artistic and creative accomplishment. Still, none regularly drew an audience comparable to those reached by broadcasters. *The Walking Dead* changed all that.

AMC's *The Walking Dead* (2010–) was the first original scripted cable series to compete with broadcast networks in terms of audience size. By 2012, *The Walking Dead* ranked as the top-rated scripted program of the night and week among audiences aged eighteen to forty-nine. Three years later, it was the most watched show of the week among those aged eighteen to thirty-four, eighteen to forty-nine, and twelve to thirty-four, and it drew 2.5 million more viewers than the second-most watched show among eighteen to forty-nine year olds, *Empire*, which itself grabbed headlines as an unexpected broadcast success.[1]

This success was substantial enough to have valuable consequences for AMC's bottom line. *The Walking Dead* drove a nearly 25 percent gain in advertising revenue during the first half of its fifth season.[2] To be fair, *The Walking Dead* was not a lone beacon. In the same years that it began reaching a broadcast-sized audience, TNT's *Rizzoli and Isles* (2010–2016) and FX's *Sons of Anarchy* (2008–2015) accomplished similar feats. These series proved that ad-supported cable could be more than a niche alternative to broadcast networks and could draw younger audiences that were more desirable to advertisers. Together, these series continued to adjust television's competitive norms separately from changes just beginning to be introduced by internet-distributed services such as Netflix.

The Walking Dead's ability to attract more viewers than broadcast series drew attention, but changes behind the scenes provided an even bigger adjustment to the business of television. *The Walking Dead* was the first series aired by AMC that was produced by AMC Productions. AMC Productions, a studio initiated to make series particularly for AMC Networks, was one among several new studios created to produce series for cable channels that included A&E Studios, Turner Original Studios, and Crown Media Productions (Hallmark). These studios joined Universal Cable Productions (USA, Syfy) and FX Productions, which were both established earlier in the 2000s.

The creation of these studios illustrates an important adjustment in how channels—and their owners—sought to profit from series. Rather than outsourcing production, and thus ownership, of the series they developed to other studios, the cable channels created their own studios to maintain ownership so that they could profit from the revenue that came from licensing these shows internationally and to internet distributors.

The stories earlier in the book illustrate why conglomerates created these studios. In the cases of *La Femme Nikita* and *Mad Men*, the lack of a co-owned studio nearly stymied the channels' ability to create these shows. USA and AMC, respectively, struggled to find a studio willing to produce the shows. As one of the earliest original scripted cable series, *La Femme Nikita* encountered the most difficulty convincing a studio that there could be any value in creating a series for a cable channel. Although ultimately a hit by many measures, the series returned much less revenue than a typical studio venture. Recall that Warner Bros. agreed to produce the series only after coproduction money filled in the deficit between what the production would cost and what USA would pay. And it was the studio that pushed to end production as soon as enough episodes for syndication sale were filmed.

Likewise, even though original scripted series were well established by the time AMC tried to find a studio to produce *Mad Men*, its many risky components deterred major studios from producing the series. The deal for Lionsgate to produce emerged only because it was a smaller studio willing to take on a project with considerably lower profit potential. Ironically, it is the revenue *Mad Men* ultimately produced for Lionsgate—despite its challenged start—that contributed to the emergence of these new studios. AMC received no share of the $1 million per episode Netflix paid Lionsgate for *Mad Men* after its run on AMC. Owning series became increasingly important for balancing revenue streams, which was the driving force behind the creation of these studios.

Cable channels began establishing their own studios in order to take advantage of the revenue opportunities provided by Netflix and other internet distributors that emerged as new secondary buyers of series. Producing, and thus owning, their own series was also important in adapting business models to profit from the changing market for international sales, which will be explained further in the next chapter.[3]

In some ways, the emergence of these studios is odd. Most of these channels were part of a conglomerate that owned a studio that produced most of the series aired by its co-owned broadcast network. For example, NBCUniversal created Universal Cable Productions despite also owning NBC Studios, which produced most of the series aired on NBC. The decision to create separate studios illustrates the persistent differences between cable and broadcast production. The skillsets and practices were not interchangeable: broadcast studios developed series targeting a broad cross section of the audience, while cable studios created distinctive content designed to resonate with a niche audience and establish a channel brand. Further, the budgets of broadcast and cable series also made the business of making cable series different enough to warrant a separate studio infrastructure.[4]

As various changes throughout television and media diminished advertising revenue, companies built on selling audiences to advertisers endeavored to diversify revenue streams. They expanded the segments of their business that relied on selling intellectual property, such as developing and licensing series. The emerging internet distributors provided a new market for licensing revenue, and U.S. channels also increased their international channel holdings. The initial success of cable in the United States in the 1980s owed considerably to the audiences that could be created around niche interests when able to aggregate them across the sizable American populace. Ever more narrow niches could be monetized when drawing from a global population. The distinctive programming emerging from many cable channels attracted niche audiences around the world.[5]

The cable studios also found self-producing allowed them to better control—and thus reduce—program costs. FX Productions achieved extraordinary success with its first venture: *It's Always Sunny in Philadelphia.* Comedies tend to have lower ratings when channels first air them but draw larger audiences than dramas for repeats. FX Productions produced comedies with per-episode budgets of $400,000–700,000, which was half the budget of a typical broadcast comedy, and as little as a quarter the budget of some.[6] Eric Schrier, president of original programming at FX Networks and FX Productions, explained that creating a studio to produce original series for FX "turned us from a dual revenue stream business into a three revenue stream business."[7]

Having *The Walking Dead* produced by co-owned AMC Productions meant that in addition to the increased advertising rates it was able to charge for its massive audience, AMC was also the beneficiary of the sale of *The Walking Dead* to the MyTV broadcast network, Netflix (U.S. and Canada), and Lovefilm (UK), among many other deals. Of course, most series do not achieve the success of *The Walking Dead. Low Winter Sun,* which AMC coproduced with Endemol in 2013, failed to draw audiences and AMC canceled it after one season. An unpopular, single-season show is difficult to sell in other markets and required AMC to write down $61 million of losses on the series.[8]

When channel-owned studios produce series, the decision to continue them for subsequent seasons is calculated differently. The next series produced by AMC—*TURN: Washington's Spies* (2014–) and *Halt and Catch Fire* (2014–2017)—drew far smaller audiences than *The Walking Dead.* AMC renewed these series for subsequent seasons, although neither attracted a fraction of the attention of earlier AMC shows.

Owning series that weren't enormous hits could still be valuable—especially if there were other audiences to reach. As the U.S. market became saturated with programs and program suppliers, cable channels increasingly looked abroad to expand their businesses. AMC purchased Chellomedia, a company that owned a wide mix of international channels in 2014.[9] The next competitive move for U.S. cable channels was taking their shows to a global audience. But could they establish audiences in other countries faster than emerging internet-distributed services—such as Netflix—that had similar global ambitions?

17 Cable Goes Global

The international market has always been an important part of the U.S. television business. The large and relatively wealthy U.S. population has allowed the television industry to afford expensive productions and recoup costs in the domestic market alone. Selling shows to channels in other countries added important revenue, but this wasn't as central a consideration in creating shows as it became by 2010. Escalating program budgets, smaller and more fragmented audiences, and the growth of serialized programs[a] that don't sell well in domestic syndication led studios to increasingly rely on international markets to cover the costs of production.

Multichannel television—and its access to hundreds of channels whether delivered by cable, satellite, or telco—arrived later to countries outside the United States. When it did, channels bought up a lot of U.S. series to fill out their schedules. But by the late 1990s, the U.S. studios began to price themselves out of many markets. These shows produced before the programming transformation initiated by original cable series didn't excite audiences outside of the United States, which made their high cost not worth the limited audiences they would draw. Channels outside the United States began developing alternative series such as workplace docusoaps in order to fill schedules with programming less expensive than the prices demanded by U.S. studios.

a. *Serialized programs* tell an ongoing story—more like a soap opera—or at least require knowledge of what happened in previous episodes. Until the 2000s, most U.S. primetime television told *episodic* stories, or stories that resolved in a single episode. Programs structured episodically were more valuable in syndication because they could be aired out of order and didn't require that audiences see every episode for them to make sense.

International tastes did not exactly mirror the American market. Police procedurals such as *CSI* perennially earned considerable revenue abroad, but increasingly ambitious dramas that ranked among the more innovative of broadcast-fare—*Desperate Housewives, Prison Break, 24*—were more popular internationally than their U.S. performance would have suggested. Likewise, the distinctive programming developed by cable channels proved popular around the globe. U.S. series rarely ranked among the ten most popular series in European countries, where audiences preferred locally produced content. The growing potential for cable series outside the United States became clear when USA's *The 4400* —a series that performed solidly among U.S. audiences, but was far from exceptional— ranked fourth in Germany and was one of the top shows in the UK in 2006.

The U.S. television industry expanded its global reach throughout the 2000s by selling shows internationally, and also by launching channels distributed on expanding satellite and cable systems. Once the U.S. market for multichannel service reached as many homes as would likely ever subscribe to the service, the only way for the U.S. cable business to grow was to expand globally.

Some cable channels developed internationally from the early days of cable. Discovery, HBO, Viacom (MTV), and Turner (CNN, Cartoon) have several international channels that date to the 1980s. Many other early ventures failed because they expanded too quickly or did not adequately take into account the different competitive norms and cultural expectations of television in other countries.[1] For many other channels, the substantial expansion in international channels occurred throughout the 2000s.

The development of international channels took different forms. Truly transnational channels—channels with consistent programming around the globe—rarely emerged. More often, only the channel name was transnational. For example, HBO developed many channels in other countries and regions—each named HBO and the country or region (HBO Brazil; HBO Nordic), but the programs and schedule of these channels were entirely separate from the U.S. channel. In many cases, Warner Bros., the parent company of U.S. HBO, was a partner rather than the sole owner of the international channels and sometimes simply licensed the channel's name. As a result, the programming often differed substantially from that of U.S. HBO.

Many cable properties spanned the globe by 2014: A&E's *Pawn Stars* reached 325 million households in more than 180 countries; *SpongeBob SquarePants* aired in more than 185 markets (the most widely distributed property in Viacom history); and Discovery brands such as Discovery Channel, Animal Planet, and TLC appeared in more than 220 countries.[2] Although it lacks a channel in the United States, Sony had more than 120 channels around the globe and production facilities in places such as Russia, where it produced localized versions of U.S. hits such as *The Nanny* and *Everybody Loves Raymond*.[3] Even broadcaster CBS launched twenty-two international channels between 2006 and 2013. The channels generated $1.1 billion in international revenue as of 2011, twice the total in 2007.[4]

In addition to a brand identity, the global channels often benefited from economies of scope. Programs developed for the U.S. channel aired with dubbing or subtitling as needed. In other cases—such as *Jersey Shore* or *Real Housewives*—channels would use a U.S. format and make a locally relevant version. Channels still incurred production costs, but they saved the costs of new development and had the benefit of knowing the concept had worked with an audience in another region.

In cases of series particularly in demand, distributing on a self-owned channel might not be the most profitable strategy. AMC sold *The Walking Dead* to Fox for international distribution on its channels because it proved more lucrative given AMC's limited international holdings. HBO often sold its programs to non-HBO channels. Its original programs were in such high demand that it was more profitable to license them to other networks—even in markets where an HBO channel existed. Though U.S. audiences know HBO as a distinctive, subscriber-supported service, its shows were regularly found on ad-supported channels elsewhere.

The international channels were often separate enterprises, designed for the specific characteristics and competitive environment of each country. All of this variation made the universe of cable competition very different around the globe, even though many of the channels and programs seemed the same. It also helps explain why cable channels created studios that would allow them to own the series they developed.

The internationalization of cable channels emerged as a business strategy, but it also raised a variety of concerns. Many had long been concerned about the consequences of exporting the values pervasive in American programming around the globe when international trade was based only on selling programs. Transplanting entire channel concepts expanded these concerns.[5]

Internet distribution was also beginning to challenge norms of selling series to channels in different countries and regions. Netflix expanded to offer service in 140 countries at the start of 2016, and by the end of the year, Amazon Video made its original series available in two hundred countries and territories.

What viewers could watch on Netflix differed considerably around the world. The content included in Netflix libraries varied based on the regions where Netflix had licensed various shows. Early Netflix original series such as *House of Cards* were not available in most international markets because Netflix had not purchased the international rights—although that practice soon changed.[6]

This chapter just grazes the surface of what was—by 2016—just the beginning of a reconfiguration of the international television business. Television businesses rooted in broadcast distribution emphasized national boundaries, but internet distribution introduced new norms as programs moved more easily across boundaries to gather global audiences. New distribution technologies require adjustments in practices for making television and strategies for monetizing it.

The anxiety about the "future of television" in the United States has minimally addressed the now global nature of these businesses. Uncertainty about the future of U.S. television is only made cloudier by the realization that this future is intertwined with global markets. Recalibration of global television business practices provides considerable opportunity and yet has unknown consequences for distributing and accessing television.

18 Watching Cable Before the Internet

Many may have viewed, or at least heard of, shows such as *The Shield, Monk,* and *Mad Men* during the first decade of the twenty-first century, but far fewer knew how the business of television changed to allow their creation. Television programming and the business behind it wasn't the only evolution in these years; the experience of viewing was also shifting in viewers' homes and before their eyes.

Digital cable made the experience of an abundance of channels the dominant experience of television; 77 percent of homes with cable had digital service by 2010. Adding the 32.2 million satellite homes, and five million receiving multichannel service from a telco, 73 percent of U.S. television households had digital service.[1] Having so many choices meant that fewer and fewer shows gathered large audiences. Homes may have had a lot of channels, but the collection of channels each household viewed varied considerably.

The change in television in this era wasn't just a matter of choice, but of image quality as well. The use of high-definition television also grew tremendously in this period.[2] Most homes didn't begin receiving high-definition signals until they upgraded to digital cable. The shift to digital consequently brought both better images and more channels.[3]

Devices for managing the choice also grew more common. Some of these changes didn't begin entering the homes of "typical" viewers until after 2010, but important groundwork for coming disruption was established in this decade. DVRs gave viewers more control over *when* they watched, which also adjusted *what* they watched. DVRs provided a self-selected library of shows ready to watch whenever desired. Such technology diminished the frequency with which viewers simply chose something because “it was on” and made managing the growing abundance of programming possible.[4] Still, in 2010, DVRs were in only 38 percent of homes. Viewers in those homes watched an average of twenty-six hours of recorded content per month, a mere 17 percent of the hours viewed.[5]

Further behind the screen, competition among cable, satellite, and telco providers evolved quietly but with profound consequences. Cable remained the dominant video provider, but its share of the market diminished. Cable reached its peak number of video subscribers in the early 2000s as satellite proved a formidable competitor in video service. Satellite subscribers nearly doubled from 2002 to 2010, growing from 17 percent of multichannel households[6] to 30 percent of those subscribing to multichannel services in 2010.[7] Much of satellite's gains came at the expense of cable, which fell from seventy-three million subscribers in 2002 to fifty-nine million in 2010.[8]

The anticipated competition created by the Telecomm Act arrived at last, though few noticed. The two major remaining wired telephone companies—Verizon and AT&T—effectively cherry-picked the markets and neighborhoods in which they upgraded wires and offered service. As a result, few homes—typically the most affluent—even knew there was an alternative to cable or satellite.

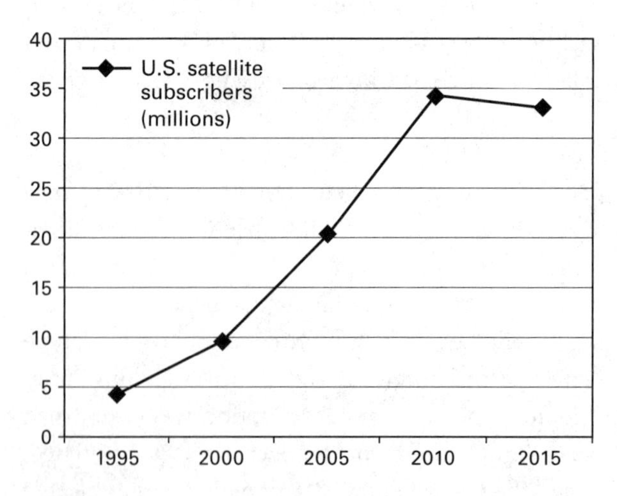

Figure 18.1
Satellite subscription growth, 2002–2015.
Source: ITVA; SNL Kagan.

Verizon launched its FiOS video service in September 2004. It achieved only one million subscribers by January of 2008 and another two million subscribers by January 2010, at which point it announced the end of its service expansion. The service was available to 12.2 million homes, but it had rewired only one-third of the areas in which it provided voice service.[9] FiOS service was available nearly exclusively in the major cities along the East Coast.

The story of AT&T's U-verse service is little better in terms of bringing competition to the video market. The service launched in seven cities in 2006 and added thirteen more in 2007, but had only 100,000 customers by September of that year. Its subscriber base grew to 2.5 million by 2005 and reached three million by 2010. AT&T acquired DirecTV in 2015. Its combination of satellite and telco distribution promised adjustments to the competitive field.

The limited accessibility to telco competition meant that there was great variation for Americans seeking multichannel service in the first decade of the twenty-first century. For example, those in the greater Washington, DC, area—incidentally the home of most regulators—could choose among FiOS service, cable, or satellite and the market was more equitably shared with 51 percent of multichannel households subscribed to cable, 29 percent to satellite, and 19 percent to telco service.[10] But few other cities had such a competitive marketplace. Nationwide, 60 percent of homes that paid for one of these services subscribed to cable, 33 percent to satellite, and 6 percent to a telco.[11]

Cable channels dominated attention to "cable" in these years. The cable provider business struggled through the early 2000s as it endeavored upon massive infrastructure upgrades and was presumed under threat from "the internet." Attention to changes in television programming and the innovation of cable channels obscured the fact that during the late 1990s and early 2000s the cable industry weathered the biggest challenge to its business and emerged with the metaphorical golden ticket. While most thought the future of television had something to do with the relative merits of *Breaking Bad* versus *NCIS*, a surprise twist developed in what many expected to be cable's last act. The cable service industry that was largely forgotten during the flurry of attention to shows produced for cable channels resurfaced at decade's end, revealing its transformation into the internet service industry.

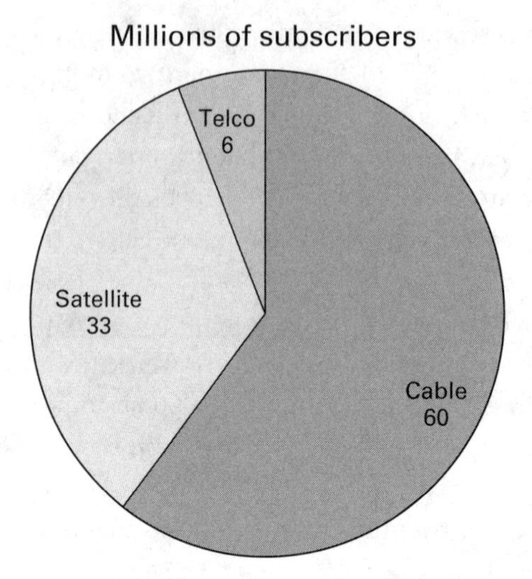

Figure 18.2
Market share of multichannel providers, 2010.
Source: SNL Kagan, 99.5 million multichannel homes.

The cable industry established a substantial advantage in incumbency before anyone realized that the future of television was intricately tied to the internet.[12] Knowing what we know now—that the majority of U.S. homes would pay for high-speed internet access by 2007 and that internet-distributed television would become a competitive part of the television industry in 2010—it is difficult to understand why many counted cable providers out.[13] Few fully understood how the landscape was about to shift and that *cable providers would become internet providers for most U.S. homes*. In many ways, it was not until viewers—whether ordinary Americans or stock analysts—experienced internet-delivered, long-form video as first made available by HBO Go and Netflix in 2010 that anyone appreciated the enormity of internet distribution for the future of television. By the time that moment arrived, the cable industry had already become the internet industry.

The now-combined cable and internet service companies did struggle with the "cable" component of their business. Their profit margins substantially decreased as channels negotiated for higher subscriber payments. An even bigger challenge for cable providers resulted from an unanticipated response from broadcasters to reassert their status. Once cable series began dismantling broadcast dominance, the broadcast networks and their conglomerate owners began to longingly eye cable's dual revenue streams. Shifts in approaches to "broadcast retransmission" payments—the arrangements local broadcast stations negotiate for carriage on cable services—evolved considerably beginning in 2007, and broadcast networks too enjoyed a dual revenue stream by 2010. These fees raised cable service providers' costs considerably. Those fees were passed to subscribers and stoked existing dissatisfaction with the high price and lack of choice in multichannel service.[14]

Although internet distribution plays little role in the first half of this story, from 2002 to 2010 cable companies laid the infrastructure groundwork upon which those services would rely. The experience of home internet changed tremendously between 2000 and 2010 even if it was not yet a significant factor in the business of television. There were 55.4 million internet households (50 percent of households) in the early 2000s. Only 29 percent of those homes had high-speed service capable of transmitting video—or 14.5 percent of all U.S. households.[15]

The majority without high-speed internet did not have a dedicated line for internet service, but rather tied up their phone lines when dialing up to check email or use the web. Users paid monthly internet service fees that allowed access for a certain number of minutes per month. It required 12.5 minutes to download a song and an average of sixteen seconds for a web page to load using dial-up service with the common 56K modem.[16] Although it was not so long ago, these features of using dial-up internet offer a valuable reminder of the different experience of internet access in the mid-2000s. This experience fueled a much lower expectation of the internet's impact than what became the norm by 2010.[17]

Increased standards for high-speed internet and the expansion in internet service to the majority of the population extended the importance of internet access so that it became integral to daily life. This was the dawning of the digital modernity that has redefined so much of the life, work, and culture of the societies into which digital infrastructures rapidly expanded. Accessing television is of course just one, arguably unsubstantial, part of that transition, but it provided an illustration of how significantly and rapidly a familiar aspect of daily life might change.

Despite these developments in the number of channels, image quality, control technologies, and the competitive field of multichannel services, the first decade of the twenty-first century was the relative calm before the storm as far as changes in how television series are made, financed, and come to reach viewers. By the time that first decade drew to an end, waves of shifting industry practices crashed more quickly, each bringing more disruption than the last.

19 Distinction Fails

Between 1996 and 2010, cable programming went from being a backwater of old broadcast series to being a source of ambitious and original storytelling. In time, cable programs matched and regularly surpassed the creative accomplishments of broadcast series.[1] Cable was no longer some lesser form of broadcast television. It became seamlessly integrated into what Americans considered television to be.

As a result, cable channels completely reversed creative workers' perception of what it meant to develop series for cable. In 2002, executives at USA and FX struggled to find talented writers and producers willing to make series for cable. Only those who had yet to develop a reputation entertained this risk. A special few leapfrogged ahead in their careers because they took a chance on the opportunity.

Within just a few years after *The Shield* and *Monk* debuted, talented creatives with established reputations struggled to get meetings with cable channel programmers because creating series for cable became synonymous with greater creative freedom. Broadcast series remained considerably more lucrative for creative talent, but the latitude available in cable's pursuit of distinction was a siren song for those with a story to tell.

The business of television changed accordingly. The growing abundance of cable shows and channels collectively ate away at broadcasters' dominance but, for the most part, broadcast shows still reached far more viewers. Eventually, mass cable hits emerged and steadily broke viewership records to suggest just how thoroughly old norms had been disrupted.

Many had already forgotten these changes in television by the end of the first decade of the twenty-first century. Internet-distributed television—the force of change driving the story in the second half of this book—

quickly disrupted this new equilibrium and adjusted far more than television programs. It is important to remember, however, how much television changed—and why it changed—before the arrival of internet-distributed television. It was not simply that broadcast networks dominated, then *The Sopranos* came along, then Netflix, and suddenly everything was different.[2]

Before moving on to the transformation wrought by internet distribution, it is worth pausing and reiterating the substantive preexisting change—and crisis—into which what was casually called *streaming* emerged. As the story so far illustrates, the dynamics of U.S. television changed profoundly even before viewers streamed a single television show. When this book's story begins in 1996, dozens of cable channels were available, but those channels brought little new or original scripted programming into viewers' homes. In the roughly fourteen years explored so far, cable channels multiplied and became plentiful sources of original programming. Of course, many channels also continued to fill much of their schedule with acquired programs. Still, an abundance of original content quickly developed as channels recognized the strategic value of distinctive programming.

By 2010, signs appeared that this abundance might tip into surplus. On its face, this might seem a good thing. Rather than the widely forecast death of television, the medium seemed to be reborn. There was more television content—more really good television series and wide ranging shows for a greater array of audiences. Some of it was truly excellent, and viewers continued to spend more hours watching it.

Pinpointing when the abundance of programming became a surplus is impossible; no particular year or set of series made such a turning point clear. Discussion of something called *peak TV* became a theme among television critics writing in the summer of 2015 after FX president John Landgraf illustrated precisely how much original scripted series production had grown. But articles opining that there was too much to watch had been circulating for a few years.

The sense that there was too much television was a personal assessment. Television critics—whose jobs had required reviewing everything—were canaries in the coal mine. Most viewers didn't feel similarly compelled to sample all new shows. But the feeling of being unable to keep up with all the shows that friends and coworkers recommended became more palpable as cable originals multiplied. Despite increased quantity and distinction, there were still many stories not being told by television.

The assertion of a surplus was easier to make in terms of business metrics. Landgraf circulated a chart illustrating the escalating number of series. In terms of the time frame considered so far, the change was notable but not yet profound. Broadcasters actually offered fewer shows in 2010 than 2002, but cable doubled its output from forty-eight to ninety-nine series, which led to a net increase of about 20 percent. Cable channels—particularly ad-supported channels—kicked into high gear with a 47 percent year-to-year jump between 2010 and 2011 and a 26 percent increase the year after.

Like a rainstorm on a flooded plain, internet-distributed television then arrived to a television industry already beginning to drown in a surplus of content. In total, the number of original scripted series grew by 226 percent in the thirteen years after the launch of *The Shield* and *Monk*. The realization that distinctive, innovative content was no longer the ticket to sure-fire success was growing evident, although many channels nevertheless continued with this strategy.

Landgraf's data underscores how broader television industry practices evolved—it wasn't just a matter of internet-distributed services adding to

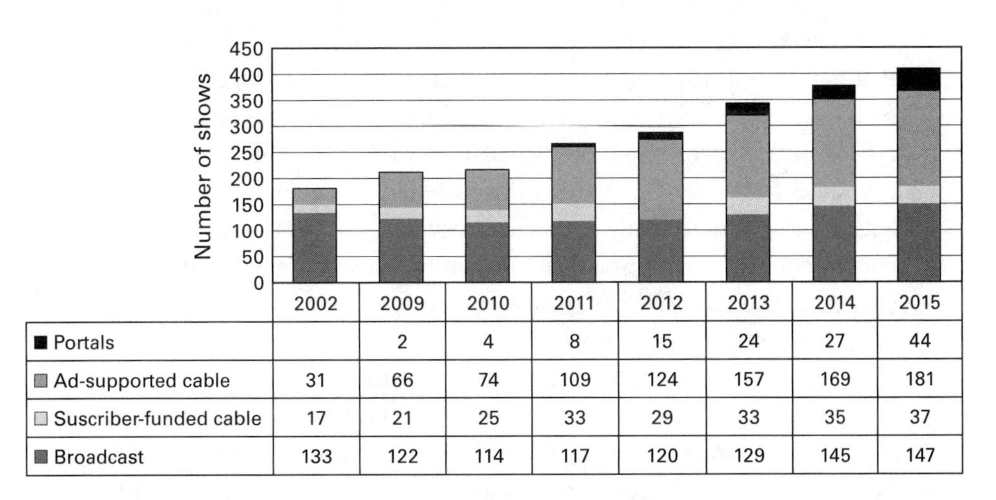

	2002	2009	2010	2011	2012	2013	2014	2015
■ Portals		2	4	8	15	24	27	44
■ Ad-supported cable	31	66	74	109	124	157	169	181
□ Suscriber-funded cable	17	21	25	33	29	33	35	37
■ Broadcast	133	122	114	117	120	129	145	147

Figure 19.1

Growth in original scripted series by provider. Portals include Amazon Video, Crackle, Hulu, Netflix, and Yahoo; excludes daytime dramas, one-episode specials, non-English language programs, and children's programs. Source: FX Networks Research.

the competition. The biggest annual increases in cable originals occurred as the broader television business shifted in 2010. Self-owned studios and greater international ownership encouraged more original production, even if having so many series made drawing audiences for the first airing difficult. The business of television had shifted by this point to be less reliant on gathering an audience to sell to advertisers.

Concern about the sustainability of such levels of production appeared as soon as peak TV was named. In the old system of television, a series' profitability could be forecast within a few years, although much of the revenue it would produce might take a decade or more to flow back to studios. Studios created this surplus of series with the assumption that later markets would produce additional revenue. But no one yet knew the new dynamics of revenue streams affected by internet-distributed services and changing international practices. A sizable readjustment in the business of television would take place when it became clear that the calculus for recouping production costs had changed. Even by the time the story told here ends in 2016 that moment had not arrived. So production soared.

Just as the shift from analog to digital cable changed the competitive field for the businesses of television, so too did the slide toward surplus. Creating "distinct" programs worked in an environment of abundant content in which distinct programming could stand out, but an environment with an abundance of distinctive programming again required new strategies. As chance would have it, in 2010 internet streaming was about to introduce technological changes that would alleviate many of the challenges viewers found with the increasingly overwhelming amount of programming. The trouble was that those solutions challenged nearly every strategy of the business of television and everything its executives thought they knew about the medium.

The surplus of programming was so pronounced that scarcity of time emerged as a real problem for television's business model. Advertiser-supported channels especially struggled. Their business model remained tied to live, or near-live, viewing. Viewers could somewhat manage the surplus with devices such as DVRs, which allowed them more flexible viewing, but much of that viewing did not "count" for advertisers. Even more disruptive to ad-supported channels were the hours of viewing that started to move to Netflix and Amazon Video. Viewers finally had tools to push back against various aspects they disliked about television viewing and the television industry.

The death of television was frequently predicted in the early years of the twenty-first century. By 2010 the situation seemed especially dire; yet paradoxically it was more hopeful, too. Many sectors of the television industry were severely challenged by these new developments, but others saw great opportunity.

Major developments of the transformation: 1996–2016

1996	1997	1999	2000	2002	2003	2004	2005	2006
Programming change								
	La Femme Nikita (1997–2001)							
	OZ (1997–2003)							
		The Sopranos (1999–2007)						
				The Shield (2002–2008)				
				Monk (2002–2009)				
					Queer Eye (2003–2007)			
					Mythbusters (2003–)			
					American Chopper (2003–)			
							The Closer (2005–2012)	

1996	1997	1999	2000	2002	2003	2004	2005	2006
Industrial change								
Telecommunications Act of 1996			Significant merger/ consolidation among cable providers		Voice over IP (cable delivering home telephone service) achieves standard availability	Verizon launches its FIOS video service	iTunes announces $1.99 downloads of TV episodes	Google buys YouTube (Oct) AT&T launches its U-verse video service

2007	2008	2009	2010	2011	2013	2014	2015	2016

The Sopranos (1999–2007)
The Shield (2002–2008)
Monk (2002–2009)
Queer Eye (2003–2007)
Mythbusters (2003–)
American Chopper (2003–)
The Closer (2005–2012)
Mad Men (2007–2015)
Jersey Shore (2009–2012)
The Walking Dead (2010–)
House of Cards (2013–)
Orange is the New Black (2013–)
Horace and Pete

2007	2008	2009	2010	2011	2013	2014	2015	2016
					The *Walking Dead* is most watched show on television VOD libraries become viable Amazon Video releases first series *Alpha House* and *Betas* (Nov)		*Game of Thrones* released simultaneously worldwide (April)	
Nielsen identifies 2 million zero-TV homes (homes using only internet-distributed services) Smartphones debut Broadcast networks begin negotiating for retransmission fees	Hulu debuts (March)		Netflix starts streaming-only service, subscriptions surge HBO Go debuts (Feb) Apple releases iPad (April) Industry worries about OTT Smart TVs come to market	Netflix announces splitting of DVD and streaming service and launch of Qwikster (Sep) Netflix reverses decision to split DVD and streaming service and launch Qwikster (Oct)	U.S. multi-channel subscrip-tions decrease for the first time Nielsen identifies 5 million zero-TV homes	OTT/cord cutting enter popular lexicon	FCC establishes net neutrality policy Half of U.S. homes have TV sets that connect sets to internet HBO Now, CBS All Access launch	

Part Two The Internet Revolutionizes Television, 2010–2016

Before strategies for dealing with surplus could be identified—or even before it was recognized as a problem—the challenge had already become significantly greater. Internet-delivered video services seemingly emerged overnight and brought with them more opportunities to watch television as well as more programming. Part two of this book traces the preliminary stages of competition between the *legacy television industry*—those companies that competed before the arrival of internet distribution—and the emergent internet-distributed sector.

The most profound adjustment wrought by internet distribution, as well as technologies such as VOD and the DVR, involved freeing television from the constraints of a schedule. These technologies allowed viewers more control over what to watch and when to watch it. Such control technologies were increasingly crucial for managing the surplus of series. Freedom from the scarcity of an enforced schedule, combined with the developing surplus of series, created a very different television environment.

Some of the earliest and most widely used internet-distributed television services—Netflix, Hulu, and HBO Go/Now—introduced a *distinctive viewing experience* to this world of distinctive programming. Internet distribution enabled viewers to experience a different way of viewing. This shift required changes in all aspects of the business of television.

Internet-distributed television arguably offered a superior experience. It was entirely divorced from a forced schedule, it often offered the same shows currently "on" television (though with some delay), services were affordable relative to cable packages, and much of it was advertising-free. Although viewers experimented with and adopted internet-distributed services gradually, it quickly became clear that companies holding on to businesses rooted in television's past would be hard-pressed to compete once viewing internet-distributed television moved to mainstream practice—no matter how distinctive the content.

Throughout the first decade of the 2000s, television and the internet were conceived of as entirely different media. The internet was part of what was called *new media* and was perceived as the perpetrator in the "death of television" narrative. Most expected that the new media of the internet would replace television. Although this is not what ultimately happened, knowing that this was the expectation at that time is crucial to understanding the emergence of particular strategies and why certain paths were taken.

A confluence of factors converged in 2010 that explains identifying this year as the starting point of this part of the story. Netflix launched its streaming-only service in 2010 for just $7.99 a month, and its subscriber base jumped 63 percent.[1] The streaming service made available an unprecedented amount of the kind of video people considered "television" when they wanted it and on almost any screen they chose.

Also that year, major technological changes freed internet-distributed content from the confines of a computer. Apple launched the iPad—soon followed by other tablets—and Netflix released an app that enabled viewing on these devices as well as mobile phones.[2] Tablets and smartphones—which launched in 2007 and were used by 27 percent of the market by 2010—led to a reimagining of viewing. Laptop sales overtook those of desktops. This made computer viewing far more flexible and allowed computers to be seen not just as tools used for "work." By 2010, smart TVs—televisions that connect to the internet—as well as devices such as Xbox and Apple TV, which connected sets lacking this feature to internet-distributed content, made it easier to experience Netflix and other internet-distributed services on living-room screens.

The moment of the long anticipated face-off between "television" and "new media" arguably arrived in 2010, but few noticed. Rather than fighting a battle to the death, legacy television and internet television distributors quietly became neighbors.

At first, a surprising symbiosis emerged: internet distributors desperately needed legacy television content to woo consumers to try their services, while legacy television realized that internet distributors could provide a new revenue stream. Internet distributors entered like wolves in sheep's clothing. Distracted by the new revenue that services such as Netflix paid to license old series, legacy television allowed internet distributors to expose audiences to a new and distinctive way of viewing. This reacculturation of viewing behavior led to the demise of television limited by broadcast and multichannel norms.

The reason an immediate battle between television and new media did not occur was the way understandings of television unexpectedly, and somewhat unexplainably, evolved. Before the advent of internet-distributed television, the possibility of television on demand was curiously difficult to imagine. Without seeing it in practice, it seemed unlikely that "television" could be distributed over the internet, or that programming watched on screens other than the living-room set could be considered "television." But when Netflix and HBO Go enabled the streaming of high-definition, full-length episodes, it was, surprisingly, understood as "television."[3] People began viewing on laptops, tablets, and mobile-phone screens—and even transferring internet-streamed video to living-room screens. It would have been difficult to explain how an episode of a television show originally produced for a legacy broadcast network or cable channel became something other than television because it was transmitted over the internet. So all of it was considered television.

An explanation of why internet-distributed video was regarded as television is related to what was most watched. Video originally created for television was the most watched internet-distributed content in these years—whether clips of *The Daily Show*, music videos on YouTube, or on-demand access to shows provided by Netflix and Hulu. Rather than "killing" the legacy television business, internet distribution expanded viewers' program choices and control over how they watched legacy-produced programming.

Some predicted an overthrow of the legacy media industries by empowered amateurs who would use internet distribution to share self-produced content. Certainly, content developed by "ordinary" people and distributed through YouTube developed as a sector of internet-distributed television, especially among young people. This sector was more supplement than substitute—at least through 2016. In these early days, though, the business potential of "amateur" video was not at all clear.

Predictions of the internet's implications for television had been dire and apocalyptic. At first, internet distribution provided ways to watch television that made a self-determined experience possible and the abundance of programming manageable. An initial period of détente emerged between the legacy television industry and the new television companies built on internet distribution from 2010 to 2014. Such détente was possible because of the injection of revenue that Netflix and Amazon offered studios in fees to license series. The fact that the earliest internet-distributed services did not compete with the legacy industry for advertisers—because they were funded by subscriber fees—also aided this symbiosis. Most of the advertiser spending that was moving to support internet-distributed services such as YouTube came from budgets like direct mail rather than out of television advertising budgets. All could coexist until someone's business model foundered.

The story of how the internet came to revolutionize television begins with the story of these early years in which legacy companies play a limited role. The last four chapters of this book explore what happens after this early period, when we arrive at what seems the beginning of a middle stage of the revolution that internet-distributed television forced on the industry. Neighborly relations dissolved once internet distributors moved into producing original series and legacy providers developed their own internet-distributed services. For a few years, though, the legacy and new television industries coexisted relatively peacefully.

21 Netflix: Diabolical Menace or Happy Accident?

It is difficult to fully acknowledge the scale of Netflix's disruption of the television industry. Of course, if it hadn't been Netflix, it would have been another company. And it is not surprising that a company *without* a legacy television business charted the path toward the business of internet-distributed television. Only an outsider had such incentive to change the industry. At the same time, however, it was extremely difficult for a newcomer to overcome the barriers erected by established companies. The tale of Netflix's initial success in mailing videos to homes, then evolving into a U.S. internet video-distribution service, and then becoming something akin to the first global television network, is an amazing business story.

Regardless of what happens to the company in coming years, Netflix should be credited as a revolutionary. Revolutionaries, however, don't always rule in place of the structures they overthrow.

Netflix launched in 1997 at the start of the transition from VHS to DVD. Although the sale and rental of movies on VHS developed into a big business that became the dominant revenue stream of the film industry, little television was distributed on VHS. VHS tapes couldn't efficiently contain the twenty-plus hours of a television season. Moreover, before distinction became a core strategy, few television series produced consumer demand for selling television on tape. The idea for Netflix thus originated well before anyone imagined that people would pay to buy television.[1]

The emergence of DVD sales as a revenue stream for television series caught many by surprise in the early 2000s. Box sets of serialized shows—as well as those produced for subscriber-funded channels such as HBO—sold surprisingly well. This consumer interest expanded how studios thought about opportunities to profit from television. More distinctive series sold better than ones that drew the largest audiences, which encouraged production of such distinctive series.

Video-on-demand services—which will be explored in depth in chapter 22—provided another precursor to the different experience of television Netflix eventually offered. Switching to digital infrastructure afforded cable systems the technological ability to provide an experience very similar to the one associated with Netflix, but many cable providers struggled to acquire content. Networks and studios were wary of eroding the audiences of the existing business and did not allow the development of significant on-demand libraries until after the potential and implications of Netflix became clear. Netflix quickly surpassed cable systems' minimal VOD offerings—mostly limited to pay-per-view movie rental and the on-demand offerings of HBO and Showtime—because it was willing to pay sizable licensing fees.

Netflix's streaming service may have been built to ensure its ascendance in the film rental market, but the capacity to stream high-quality, long-form video lent itself at least as much, if not more, to television distribution. Netflix likely came to realize the demand for television content on the basis of data gleaned from its DVD-by-mail business. Or maybe it was just a hunch that if it provided a first-rate search and screening environment, even those who didn't think they wanted to stream television would quickly come to prefer the experience.

Of course streaming films was—and remains—an important part of Netflix's business. But as the company pivoted from DVD-by-mail to streaming, the core of its business shifted from film to television. Netflix remains in many different businesses: DVD-by-mail, streaming films, streaming television. What is important here is that by 2016, the split in viewing was 70 percent television, 30 percent film.[2]

Netflix's television streaming began slowly. Its first deals were for "library" or "back catalog" content. Studios were wary of making deals that might risk the conventional syndication revenue streams of current hits. Licensing back catalog content was a low risk to studios because these series were off the air and not in demand. Netflix's small subscriber base in the days of the earliest deals—under ten million through 2008—also posed minimal threat. And in many cases, the licensing deals with Netflix were a gift for studios. They often involved content that was just sitting in a vault with no prospect of earning revenue. Netflix paid considerable licensing fees, and many of the early deals were low-risk, short-term experiments with a limited subscriber base.

Several studios learned that the initial deals, in the words of one executive, "were like a free sample of crack." Studios excitedly revised profit forecasts and believed their libraries held new riches. But when it came time to renew deals, Netflix was less forthcoming with licensing fees and more particular about what it wanted. Likely armed with a treasure trove of use data, Netflix had a far better ability to value content than did the studios. And although Netflix certainly still needed content, once the studios sampled this revenue stream, they quickly became as dependent on the relationship.

Netflix included some television content at its streaming launch in 2007. This is a less significant date than 2010, because streaming remained such an early-adopter phenomenon that it wasn't particularly threatening to—or even much noticed by—legacy media until then. This assessment changed quickly over the next few years, especially by the end of 2010, when Netflix expanded its subscribers to twenty million after counting only 12.2 million at the end of 2009.[3]

By 2010, the doubting naysayers who had thought little of Netflix quickly began changing their tune. Despite oft-made assertions, Netflix was not the embodiment of "new media" come to kill television. Its service

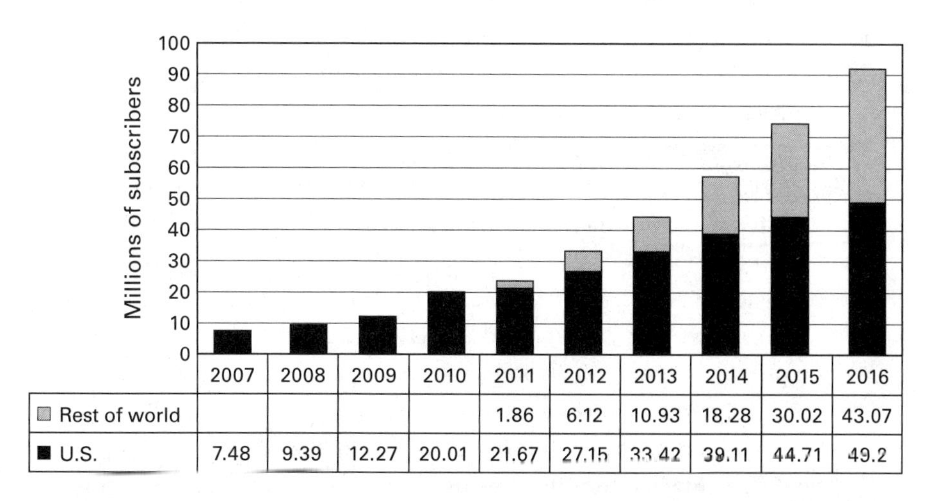

	2007	2008	2009	2010	2011	2012	2013	2014	2015	2016
Rest of world					1.86	6.12	10.93	18.28	30.02	43.07
U.S.	7.48	9.39	12.27	20.01	21.67	27.15	33.42	39.11	44.71	49.2

Figure 21.1
Netflix subscriber growth 2007–2016, millions of subscribers.
Source: Netflix.

was a supplement to rather than a substitute for cable. Its streaming library included many titles, but the "peak TV" phenomenon was just beginning, so there wasn't yet significant demand for back catalog content. Netflix's lack of new content—it typically couldn't offer series until a year after they aired—also made it an unlikely replacement for cable subscriptions. Nonetheless, Netflix clearly provided an affordable supplement that might encourage viewers to scale back their cable service packages, and by offering viewers access to a library—as opposed to offering programs in a schedule—could make it a stronger proposition in time.[4]

Netflix's effect on the television industry was simultaneously subtle and profound. More than any other development by 2010, Netflix was responsible for changing how people watched television. Cable may have disrupted broadcast norms with its strategy of distinction, but only by expanding the range of stories television could profitably tell. Netflix challenged both broadcasting and cable by enabling completely new viewing practices.

Netflix's introduction of the practice of "bingeing" was the most frequently discussed change in viewing. This term was first used to describe hours-long marathons of viewing, but most people didn't use the service this way—or at least did so rarely. More important, Netflix was the first widely used service that allowed viewers to experience television outside a network-determined schedule. It enabled viewers to watch episodes in quick succession—maybe not a "binge," but not with a week gap between episodes either. It was viewers' gateway into choosing how to view rather than having a viewing schedule forced upon them.

For the most part, viewers *loved* Netflix, and in 2010 and 2011 the service was widely celebrated. Netflix's library was particularly rich in these years—offering current season and select library content. But providing a distinctive viewing experience was not a competitive advantage Netflix could maintain in the long term.[5]

The legacy television industry took note as Netflix's viability—and arguably, success—became clear. Netflix became a legitimate competitor built on the programs of the legacy industry. As Netflix's share of the attention economy grew, the symbiotic relationship between legacy television and the first major internet distributor began to break down.

The disadvantage of not owning any content of its own led to Netflix's next pivot—a shift from being a mere distributor of content to being a creator as well. Netflix's success was based on a supply of programs that could dry up if the studios developed their own distribution services and stopped selling. To compete, the service would need original and exclusive programming. Much like the cable channels in the mid-1990s, Netflix needed to offer distinctive programming unavailable elsewhere.

Netflix's first U.S. exclusive series, *Lilyhammer*, debuted quietly in January 2012 (the series was produced for Norway's NRK1; Netflix purchased exclusive U.S. rights). By that point most attention focused on the higher-profile debut of *House of Cards,* announced in March 2011 and scheduled for February 2013, or Netflix's commissioning of a new season of *Arrested Development,* due in May 2013 (discussed more in chapter 25).

Netflix's first slate of original productions—*House of Cards, Orange Is the New Black,* and season five of *Arrested Development*—established the service as a source of competitive original series. By the end of 2013, Netflix's U.S. streaming subscribership surpassed thirty million and the subscriber base of HBO. Popular perception of Netflix continued to regard it as something different or separate from television. Its distribution over the internet rather than through cable service providers was really its only substantive difference from HBO, and HBO's development of internet-delivered HBO Go in 2010 made even that distinction tenuous.[6]

Of course, YouTube and Hulu also developed successful internet-distributed television services in these years. Netflix, however, was by far the most threatening. In 2017, Nielsen data showed Netflix accounted for 46 percent of time spent streaming, followed by YouTube (15 percent), Amazon (8 percent) and Hulu (4 percent).[7] For context, research on television use in 2017 reported that only 39 percent of television viewing remained "old-fashioned" live viewing, while streaming encompassed 24 percent, DVR recordings accounted for 16 percent, VOD, 7 percent, downloaded video, 6 percent, and the remaining 9 percent classified as "other."[8]

Although YouTube garnered a lot of viewers, its reliance on content produced outside of the legacy television industry made it seem less of a threat, or at least one of a different nature.

The significant discrepancies between the norms of Netflix and those of broadcast and cable channels made it difficult to be certain why Netflix was proving so successful. One analyst estimated that the twenty-nine billion hours of Netflix viewed by U.S. subscribers in 2015 amounted to the equivalent of 6 percent of U.S. television viewing.[9] Another analyst noted that this placed Netflix sixth among U.S. conglomerate groups in total minutes of content delivered per month.[10] Understanding Netflix's quick attraction of subscribers was complicated—a matter of both its content and the viewing experience it offered. Research about why people subscribe to Netflix found 82 percent did so because of the "convenience of on-demand streaming programming," 67 percent because it is "cost-effective," and 54 percent because of its "broad streaming content library."[11] In other words, Netflix's series and movies are only the third most compelling reason for subscription, which suggests the importance of the distinctive viewing experience. Filling out the motivations, 50 percent subscribe because of the ability to watch on various devices, 37 percent for kids programming, and 23 percent for original programming.

During these early years of streaming video, Netflix helped reveal some of the differences between what technologies could do and what viewers actually wanted to do with them. Even though tablets and phones introduced unprecedented portability and lots of talk of "mobile TV," studies found that 82 percent of tablet and 64 percent of mobile phone video streaming was done *in* the home.[12] Technologies and internet distribution made out-of-home viewing possible, but this wasn't a use viewers prioritized.

Netflix's entrance of a multibillion-dollar, decades-old industry, not to mention the considerable change it quickly forced on incumbents, is an extraordinary accomplishment. Such an achievement may well be unprecedented. Notably, this never would have happened had Netflix been deploying content over the same distribution infrastructure as broadcast networks and cable channels. This is a case of the opportunities of a new technology making it possible to circumvent the barriers of entry to an established industry.

Netflix's broader influence also derives from the revenue stream it introduced to the studios at a time when business models were threatened. These deals didn't remain as rich as when they started, but they were helpful reassurance that internet distribution and the emergence of new competitors might not completely destroy the legacy business. Netflix's reliance on content created by the industry was welcomed given forecasts predicting that amateur distributed video on YouTube would replace the legacy creators. Licensing fees paid by services such as Netflix accounted for 5 percent of the operating income on average across the big content conglomerates as early as 2013.[13]

And there were several aspects of the business that Netflix did not disrupt—at least initially. When the company pivoted toward identifying more as a television channel than video rental service, it largely adopted standard industry practices for creating original series. It established deals with studios that were very similar to those common between studios and networks or channels.

But Netflix's business quickly became defined by the three attributes that most distinguished it from the legacy television business: on-demand access, subscriber funding, and global reach. Netflix wasn't filling a schedule; it was building a library. Its aims were better served by establishing deals that would allow it to be the sole global distributor of content for a long period of time—if not perpetuity. Such deals meant that studios would not be able to resell series made for Netflix to other buyers, so higher licensing fees were needed to cover costs and provide all the profit the studio might expect to receive.

Instead of the conventional practice of paying a studio part of the costs for making a series, but allowing the studio to retain ownership and earn additional revenue by reselling the series, Netflix began using a model of paying full production costs plus a percentage—called *cost plus*. The additional percentage was rumored to be as high as 90 percent in the case of prestige projects, but closer to 25 percent for most. This meant that if a season of episodes cost $30 million ($3 million times ten episodes), Netflix would pay $57 million ($30 million cost times 1.9 for a 90 percent "plus"). The chances of a series earning more than cost plus 90 percent weren't great, so such a deal would reward a studio well without incurring risk. However, it prevented the possibility of additional profits and the exceptional riches likely in royalties earned for very successful series. Series from unproven talent or more experimental series earned considerably less, but still covered costs with the initial fee.

Through the first phase of internet-distributed television, Netflix proved to be as much friend as foe to the legacy industry. Netflix capitalized on being perceived as a "new tech" company rather than a legacy media business, although it was as much the latter as the former. Until it could amass a library of self-owned content, Netflix's future depended significantly on legacy entities that were now evolving their businesses for the emerging competitive environment.

22 Over the Top of What?

Any industry is full of curious acronyms, but one of the oddest to emerge from the cable service industry was *OTT*. Standing for "over the top," OTT refers to delivering television by using internet protocols instead of conventional radio frequencies. The particular panic over OTT resulted from the fear among cable service providers that subscribers would choose to drop cable video subscriptions en masse and instead satisfy their video needs by watching internet-distributed video.

The term was perplexing, though—over the top of what? The cable box? OTT was an imprecise designation, and its imprecision obscured that over the top actually meant, quite simply, internet distributed. The panicked forecasts of internet distribution as a destroyer of cable reanimated the television-versus-new-media death match rhetoric. Such forecasts also often overlooked the very important point that, for most households, cable providers also delivered internet service. This was too significant an economic detail to overlook in predicting "cable's" coming demise.[1]

The first mentions of OTT appear in industry trade publications as early as 2005. One of the first uses is in a quotation taken from an investment banking firm report that identifies companies such as "Apple, Skype, and Vonage as companies that have built 'strong business models based on internet protocol technology, *riding over the top* of networks built and managed by cable and telecom service providers.'"[2] At this early point, most attention was on services that delivered content to portable video devices such as newly, video-enabled iPods (which smartphones would soon make unnecessary). The content going over the top of internet connections was delivered to and typically downloaded onto these devices for later viewing.[3] Limitations in internet speeds and compression technologies made live streams of long-form video uncommon in 2005.[4]

The etymology of OTT acknowledges video being delivered by services that don't also provide the infrastructure that carries it—they send their video "over the top" of an infrastructure built and maintained by someone else. Previously, the industries that delivered video—first broadcast stations and then cable service providers—constructed and maintained the facilities that allowed for video transmission. OTT services such as Netflix rode along a road paved and maintained by someone else.

This feature of internet distribution is not particular to video. Indeed, it is a core reason why the internet has been so disruptive to many industries. One could argue that cable channels too travel "over the top" of an infrastructure maintained by cable and satellite providers. The key difference is that the channels and cable service providers have agreed-upon terms of service for that carriage. Money is exchanged and distribution is mostly contained within the cable providers' infrastructure. But that set of arrangements emerged because cable infrastructure has no value without the channels—cable providers needed cable channels to gain subscribers. Internet subscribers derive a range of benefits from internet service that encouraged their desire for the service—for which they pay significant monthly fees—aside from accessing television.

At first, published discussions of over-the-top video remained largely isolated to industry discussions of technology and reports by business analysts of coming changes. By 2010, however, a feeling of panic emerged and began to overwhelm the sobriety of earlier discussions. This feeling was tied to the emergence of another term—*cord cutting*—which appeared in just fifteen trade press articles in 2008, grew to sixty-four in 2009, and then more than doubled to 158 in 2010. Cord cutting describes the practice of canceling cable video service. Over-the-top alarm reached full swing within industry circles in 2010, although it remained mostly outside the awareness of the general populace.

The discussion of internet-distributed video from 2010 through 2014 takes on a "sky is falling" tone but, at this point, the sky remained firmly intact. Much of the coverage simply replaced the previous death-of-television narrative with one predicting the "death of cable." Such simplistic assessments overlooked the complexity of the different industries involved, especially the distinction between cable channels and cable providers. The forecasts of its demise were premature because those forewarning about the end of cable failed to acknowledge that "cable" service companies derived considerable revenue from providing internet service.

Evidence of cord cutting remained minimal, and many of the "studies" being reported relied on asking people about their plans and intentions rather than evidence of actual behavior. Subscriber frustration with the cable industry was rampant. This might have led to an overreporting of intentions to end cable service. Others may have indeed intended to cancel their cable service without realizing that it would mean losing content not otherwise available.

The panic over internet distribution developed in 2010 because internet distribution of television had at last become viable. Internet-distributed video had been technologically possible earlier, but the experience was decidedly inferior, content was very limited, and desktop computer screens were assumed to be the only alternatives to watching on a television set. In 2010, however, the type of content many wanted to watch became available, and new devices for viewing emerged that broke the binary understanding of screen technologies as either television or computer. Still, frequency of cord cutting was not commensurate with the apocalyptic predictions.

Focusing on "over the top" and cord cutting obscured the real and important transition that was taking place. While industry summits, blog posts, and tech articles continued to warn of television's destruction at the hands of OTT, viewing OTT as a foreign invader rather than a new distribution mechanism stressed only the challenge and hid its opportunity.

OTT wasn't about to kill television. The term was really just a poor description for a new way of distributing it. Like broadcast waves, cable wire, and satellite beam before it, internet protocols now provided an alternative way to distribute video. Internet protocols allowed capabilities impossible for previous distribution technologies. In particular, internet distribution didn't have to send the same show to everyone at the same time, so it allowed viewers to access video on demand. These capabilities were indeed transformative—arguably even revolutionary.

Most of the normal practices of broadcasters and cable channels resulted from the abilities and limitations of earlier distribution technologies, particularly that they had only the capacity to send one show to everyone. This limitation led to norms such as the organization of television into a schedule of shows offered by channels. (Actually, this had been a requirement of broadcast television; cable simply adopted it.)

How else could television be organized? What business models might better fit the capabilities of internet distribution? These were big questions requiring bold thought leadership. But the conversation about the future of television mostly focused on how many millennials were watching YouTube and opining about whether Netflix would destroy HBO.

Most viewers knew little of the tempest of OTT that monopolized industry press and executives in the first half of the 2010s. Technically inclined early adopters—many of whom were facing escalating cable bills—tried out the emerging internet-distributed ways of accessing television. Although Netflix grew increasingly mainstream and YouTube ubiquitous, "going over the top" remained isolated to a niche group during these years. The growing use of internet-distributed services supplemented legacy television use, but did not end or threaten multichannel dominance.

The most radical change introduced by internet distribution was freeing television from a schedule. The "immediacy" or "liveness" that had been believed integral to the experience of television was revealed to be a byproduct of its previous distribution technologies. Internet and other digital technologies allowed viewers to access television at their command—as had occurred with print and audio media. Viewers rejoiced, but television's business models now needed to catch up.

23 TV Whenever, Wherever

Netflix's considerable role in ushering in new ways to experience television derived as much from Netflix's business as from what its business prompted others to do in response. Concern about Netflix spurred cable providers to innovate and think creatively beyond the experience broadcast television offered.

Cable providers had been capable of offering a different viewing experience since they rebuilt their distribution infrastructure in the late 1990s. Digital cable—especially by 2010—was technologically very similar to internet distribution. Cable service providers increasingly sent their video signals using internet protocol. Having consumers imagine that the video they received as part of their cable subscription and internet service were two entirely different things was more a strategy for billing than a matter of technology.

But as we have seen, cable channels aimed for parity with broadcasters—to be the equivalent of "TV." Doing so—at least until 2010—meant replicating what broadcasters did rather than taking advantage of digital cable's ability to be less schedule-bound and offer television on demand. As internet-distributed television grew mainstream, cable channels needed to reconsider the competitive landscape.

The evolution and revolution of television was not simply a matter of technology. The ability to watch on demand was meaningless without programming viewers wanted to watch, and providers couldn't make on-demand programming available unless there was a way to earn revenue from it. Cable had the technology, but for roughly the first decade that cable could offer video on demand, it couldn't secure—or decided it couldn't afford—the audience-desired programs that would have made cable providers, instead of Netflix, the innovators.[1]

Like Netflix, cable providers were distributors, not creators, of content. They had no self-owned content to offer and, unlike Netflix, they weren't willing to pay to acquire it. Viewers desired the shows offered as part of networks' and channels' regular schedules. The studios and networks that owned these programs or rights to their initial airing sought to protect their businesses by preventing the development of other ways of accessing content. By 2005, networks and channels were facing the first significant challenge to their schedule-based business from the increasing availability of DVRs. The legacy industry remained deeply invested in maintaining the schedule because it preserved the legacy businesses based on distribution and the revenue model of selling and reselling television shows. On-demand services from cable service providers threatened this revenue. Studios and networks consequently resisted making these series available for VOD access.

Cable service providers could have bought original programming to create on-demand libraries. Until Netflix, however, there wasn't significant consumer clamor for on-demand capability—or even if viewers did want to watch that way, it seemed impossible. Given this lack of demand, cable service providers doubted that the amount they could earn from selling advertising or increasing subscription fees would compensate for the cost of programming. Consequently, VOD mostly went unused until internet-distributed competitors emerged and viewers became acculturated to on-demand use.

VOD foundered for several years. For its first decade, it consisted mostly of pay-per-view movies, material such as "making of" shorts, or branded entertainment—such as profiles of new cars produced by car manufacturers—that redefined the norms of infomercials. The technological capability was in place, but it lacked desirable programming.[2]

As is the case throughout this story of how cable changed television, subscriber- and ad-supported cable channels had very different approaches to video on demand. Free from negotiations with advertisers and unconcerned about how and when viewers watched, subscriber-supported channels moved into VOD early and aggressively. HBO launched the first on-demand service on a Time Warner system in South Carolina in 2001, and Showtime followed in 2002.[3] The on-demand offerings helped subscribers derive more value by making it seem that the service provided more content. Making their library of programming more readily available increased the likelihood that viewers would find something they wanted to watch from the service without significantly expanding programming costs.

Beyond the subscriber-funded channels, VOD arrived like a thief in the night over a decade later. Until 2013, most viewers had little reason to consult their VOD menu unless looking to watch—and pay for—a movie.[4]

The transition of advertiser-supported VOD from infomercials to desirable shows happened slowly and without much fanfare. Change was gradual because each cable service provider (Comcast, Charter, Time Warner) negotiated content for VOD individually, so offerings were wildly variable across the country. Moreover, each cable service provider negotiated separately with every content holder (Disney, 20th Century Fox, NBC-Universal, Viacom, Time Warner)—typically at the time carriage fees were renegotiated—which created further inconsistency in VOD availability. As a result, there isn't a specific date or event that marks when VOD systems began to be populated with current broadcast and cable series.

By 2013, many more prime-time cable and broadcast programs were available for free VOD access, but the availability of programs varied widely. Some channels allowed the most recent episodes to be accessible on VOD systems, while another channel might restrict availability to only the most recent five episodes. Sometimes shows would be available right after they aired; in other cases, there was a week delay. The duration of their availability varied similarly. Such inconsistency mirrored the slow experimentation with VOD availability and led to limited VOD use as content gradually trickled into cable providers' libraries and became more reliably available.

The deals to finally make current programs available on demand took place in tense negotiations between cable service providers and channels/studios as early as 2010 and resulted from broader changes in the industry.[5] Cable service providers had lacked a strong negotiating position until broadcasters began seeking retransmission fees in exchange for including broadcast networks on cable systems around 2007. Cable service providers were able to make VOD access part of the negotiation—especially after both parties saw how Netflix had developed in the market and recognized the need for countermeasures.

The turning point that most clearly signaled the arrival of VOD occurred in March 2013, when Comcast premiered its Watchathon week in which all subscribers could access its full on-demand library. The Watchathon included access to HBO and Showtime On Demand, even for those who didn't subscribe to these channels.[6] Comcast boasted the availability of 3,500 episodes from thirty channels and networks. Subscribers to the service's most basic tiers also gained access to Comcast's Xfinity On Demand service, which was not normally included with their service package and included the service's free VOD television offerings. The Watchathon promotion drove viewers—who may not have realized the extent of the offerings—to the VOD menu to discover its new depth. Comcast reported a 40 percent jump in VOD use between 2011 and 2013.[7]

For viewers trying to make sense of the increasing nonlinear options, VOD and Netflix were more complementary than competitive. Current seasons were rarely available on Netflix (except for its original and exclusive series), while the availability of previous seasons varied considerably on VOD. Some providers did endeavor to develop substantial back-catalog offerings; Comcast's Xfinity On Demand was clearly designed to provide viewers with a Netflix-like experience, although it wasn't nearly as easy to navigate until it upgraded its cable box and remote-control technologies.[8]

The bigger difference between VOD and Netflix was that VOD attempted to maintain U.S. television's ad-supported roots. VOD advertising practices varied substantially. Initially, the only way viewers "counted" in the sale of their attention to advertisers was if they saw the episode and its advertisements within three days of the original airing. Channels would often remove commercials after the third day and include just one or two channel promotions in the ad breaks because they wouldn't be paid for ad viewing. For example, an episode of USA's *Burn Notice* might include a promo for another USA show in each ad break and nothing else. This increased the value proposition of VOD for viewers, especially in comparison to channels that maintained the full ad load and disabled fast-forward functionality.[9]

Another impediment for VOD that marked its difference from Netflix during these years was the poor interface and search capabilities common among cable providers. Though Comcast began introducing its vastly improved X1 cable box, which more closely approximated a Netflix-style interface, in competitive markets in 2011, many VOD users struggled with unwieldy interfaces well into 2015. Part of what so profoundly pleased Netflix users was the ease of its graphic interface and the sophistication of its recommendation algorithm. Interfaces offered by cable systems remained far less attractive: they often required viewers to click through lengthy text menus, provided limited information about content, and were generally unintuitive. For example, when Comcast initially made multiple seasons of a series available through Xfinity, it organized episodes of television series in alphabetical order rather than by season. In comparison, Netflix's interface could remember where a viewer left off and quickly resume viewing. It took cable providers' systems several years to offer such advanced functionality.

By 2014, U.S. viewers made 8.8 billion on-demand requests, the great majority of which were for free on demand (77 percent), compared to just 20 percent for subscription on demand (HBO, Showtime), and less than 3 percent for transaction on demand (pay per view).[10] The future of VOD became uncertain when internet-distributed television services expanded precipitously in 2015. Many more content services began to compete with the bundles of content cable providers offered.

Cable service providers' other line of attack in competing with internet-distributed services was to follow Netflix's lead and evolve their services from a fixed-screen to a mobile-screen environment. Dubbed TV Everywhere, in 2009 multichannel service providers undertook an industry-wide initiative to create easy authentication that would allow subscribers to access the channels and on-demand content available on their living-room screen, tablets, computers, and smartphones in and out of their homes. Channels did not always have full digital rights to all the content they aired—so not all programs were available—but the TV Everywhere initiative was a sign that the industry was not complacent during this time of change.

TV Everywhere aimed to offer the same location and screen flexibility of the internet-distributed services.[11] If VOD allowed cable subscribers more flexibility in *when* they viewed, TV Everywhere enhanced flexibility in *where* to view. TV Everywhere was a notable effort by the industry, even if many viewers found it cumbersome and many reported knowing nothing about it. Enabling viewers to access content on the array of screens that had emerged by 2010 likely helped prevent cord cutting from being more prevalent. The initiative was forward thinking—especially the decision to make it a value added service instead of a new fee—but the industry struggled to make the log-in process easy. It remained unwieldy in comparison with

Netflix and exhibited significant inconsistency in available content. This fueled uncertainty about the product and its capabilities.[12]

The future of cable became a matter of discussion as internet-distributed services multiplied. These discussions rarely acknowledged that the two sectors of cable businesses—service providers and channels—were positioned very differently. Cable service providers were now internet service providers as well. Internet service had higher profit margins and these companies often had little or no competition in providing internet to the home, compared with the very different situation of cable channels.

24 Cable under Pressure

Coming right after the loud—and ultimately false—predictions of the internet's assassination of television, it was difficult to interpret the anxiety over the cord cutting that was predicted to usher in cable's demise. The panic initially seemed a red herring. Rates of cord cutting were minimal and the cable service industry provided the internet service for most of its video homes as well as those that canceled cable service. The economic well-being of these companies was not in peril. The transition to internet-distributed video was even potentially beneficial because the companies had reinvented themselves as internet providers. The profit margins on internet service were much higher than on video because of the increasingly higher fees demanded by cable channels.

The emergence of internet-distributed television began to reveal differences between cable and satellite providers. Satellite services could not provide a competitive internet service, so having subscribers go "over the top" was a greater danger for satellite companies. The different consequences for cable and satellite video providers were rarely discussed in the panic. Nevertheless, until 2017, the satellite services did not experience legions of cord cutters either.

Competition from internet-distributed services also had different consequences for large and small cable service providers. Giant providers such as Comcast, Time Warner, and Cablevision dominated the U.S. cable business, but roughly 10 percent of Americans received cable service from small providers that counted subscribers in the hundreds or tens of thousands.[1] These homes tended to be in more rural areas, and if these services hadn't been acquired by one of the large companies, it was because they were perceived as not adequately lucrative.

Lacking the scale of the giants, these smaller cable providers increasingly struggled in their fee renegotiations with channels. Blackouts—taking a channel off a cable system—were channels' trump card when negotiations were unsuccessful. Channels cared little about blackouts on these small services as they amounted to a small percentage of homes reached. The small providers had little leverage in negotiating with channels and were forced to pay more than large providers. They had no choice but to pass those costs along to subscribers who grew ever more frustrated with the rising bills.

As strained negotiations leading to blackouts grew more common and the margins on the video business kept shrinking, some small cable providers began exiting the cable business and transitioned to being just internet providers. As a result, small telecom companies serving 53,000 subscribers stopped providing video service or closed down between 2008 and 2014.[2]

Those trying to maintain business began taking the extraordinary step of dropping major channels permanently when rates could not be agreed upon. In 2014, sixty small cable service providers dropped Viacom channels such as Nickelodeon, MTV, and Comedy Central after they were unable to resolve a channel fee renegotiation. Though many expected that this was simply a typical blackout, several of the companies acquired other, cheaper, channels in the place of Viacom's offerings.

Small operators had given in to channels' demands in the past because they were afraid of losing subscribers if "must-have" channels disappeared. Given the increasing content available from internet-distributed services and general program abundance, such fears began to prove less warranted. Suddenlink, the largest of the providers that permanently dropped the channels, lost 3.6 percent of its subscribers in subsequent quarters. That loss wasn't much different from the losses of other providers that maintained Viacom channels during that time.[3] As a result, Viacom lost distribution to roughly 2.3 million households.[4]

Such developments revealed the different consequences of internet distribution for cable providers and cable channels, as well as for the largest and smallest cable providers. Many analysts focused on how many subscribers cable *providers* gained or lost as a key measure of cord cutting. An equally important measure for channels was how many homes their channel reached. Undeniable concern emerged as major channels saw the number of homes in which they were available decline by double-digit percentages.[5] Such drops, especially of channels in the basic tier, were a function not only of cord cutting, but also of small cable systems not being able to afford the rates demanded by channels.

A more nuanced picture of the peril posed by internet distribution slowly emerged. Cable service providers still had not experienced an exodus of video subscribers. Concern nevertheless continued as it became clear that a distinctive demographic of viewers—young people—was cutting multichannel service. Rather than merely delaying multichannel subscription, analysts suggested that these young people might never subscribe to a multichannel video package, earning them the designation *cord nevers*.

Young people had been at the forefront of internet-distributed video use from the outset. Many came of age on a diet of short clips of amateur content when this was the only internet-distributed video available. Yet it seemed reasonable to expect that this practice would be replaced by conventional viewing behavior as these young people entered the workforce and began families. Before they could pass through those life stages, however, long-form, professional, internet-distributed video became abundant and made realistic the concerns about a generation of cord nevers. Still,

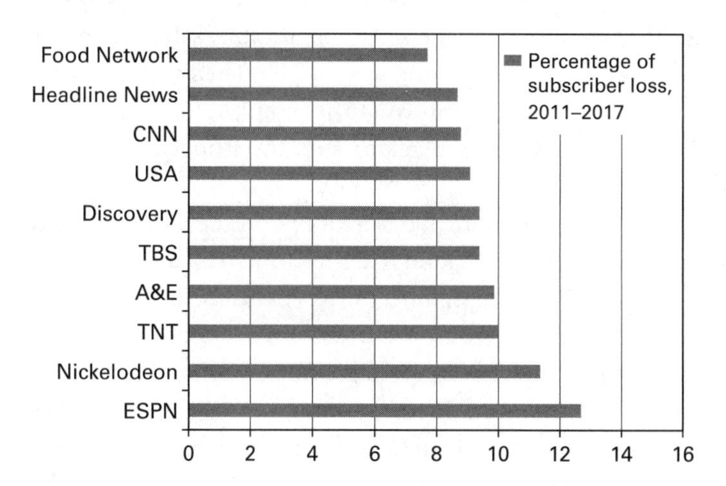

Figure 24.1
Cable subscriber losses, 2011–2017.
Source: Compiled from data in "Net Worth," *Variety,* March 21, 2017; Shalini Ramachandran and Joe Flint, "ESPN Tightens Its Belt as Pressure Mounts," *Wall Street Journal,* July 9, 2015.

2013 was the first time that U.S. subscriptions to multichannel services shrank over a full-year period. The net loss was only 251,000 subscribers, or 0.025 percent of the 100.9 million subscribing households.[6] These losses soon increased; at the end of 2016, SNL Kagan reported that the number of homes with multichannel service had dropped to 94.7 million.[7]

Beyond the behavior of the younger demographic, a phenomenon called *cord shaving* emerged as a response to internet distribution that was more common than cord cutting. Cord shavers reduce their cable package to a smaller tier of channels and augment it with internet-distributed services, for example, dropping HBO and Showtime from a cable package and instead subscribing to Netflix. By 2014, 26 percent of households reported trimming cable packages, although only 5 percent reported finding internet-distributed offerings an adequate replacement.[8] By this point, terms such as over the top and cord cutting began crossing over into mainstream publications.

An important component of such behavior was viewer frustration with the packages and prices of cable services. The arrival of internet-distributed television may have been a technological innovation, but one of its greatest effects was to introduce competitive marketplace dynamics. A lack of competition operated because viewers had such limited choice *in* providers and limited choice *from* those providers. Although cable, satellite, and, in some places, telcos competed in offering service, all offered the same large bundles of channels. The similarity in what these services offered resulted from the limited number of content creators and the stringent deals they forced on providers.

Many blamed their providers for the lack of options beyond expensive tiers of many channels, but the situation had been caused by the companies that own the channels. The cable channel owners basically held their most desired channels as ransom to secure good terms for lesser-desired channels. For example, Viacom might use cable providers' desire to have Nickelodeon to gain carriage for its other channels such as Palladia (rebranded as MTV Live in 2016) and VH1 on the basic tier. Being on the basic tier enabled the channel to reach the most homes, thus allowing those cable channels the widest possible advertising base.

The demand for increasing fees from many channels worsened the situation caused by unwieldy cable bundles. Indeed, many channels spent considerably more to create distinctive original programming throughout the early 2000s, and their spending warranted higher subscriber fees. The problem was that most channels were drawing only two million viewers to costly shows such as *The Shield* or *Mad Men*, but all ninety million households that could access the channel were paying the fees. Increasing subscriber fees was precisely how a show such as *Mad Men* became so valuable for AMC.

The increase in the monthly payment any channel achieved might have been a matter of pennies, but these incremental increases across the many channels in bloated basic tiers increased subscribers' fees to a point that led subscribers from frustration to anger. And the new fees demanded by broadcasters were sudden and in the range of dollars rather than dimes.

The shift from a competitive environment of abundant content to one of surplus also changed the value proposition. Additional content no longer added value; once viewers had more content than they could consume there was no value in more.

The threat of viewers shaving cable packages and accessing video from internet distributors was of greater consequence for channels than for the service providers. A lost subscriber hurt a channel in two ways: the loss of one monthly subscriber fee from the cable service and the loss of one household that could be sold to advertisers. By 2014, viewers were opting for lower-priced bundles to an extent that decreased the distribution of the forty most widely available channels by 3 percent.[9] This may seem insignificant, but revenue forecasts didn't imagine that distribution would shrink. When this started happening—whether because small cable providers dropped channels that demanded exorbitant fees, viewers shaved cable packages, or new youthful subscribers failed to replace to old ones—the scale of crisis became clear.

25 *Game of Thrones* Introduces the Global Blockbuster

Two key—and contradictory—programming phenomena emerged as the early years of internet-distributed television enabled new possibilities for global and independent television. The first was the creation of a global blockbuster—a "must see" series available simultaneously around the world. The second, which we will explore in the next chapter, involves the creation and circulation of a professional-level television series independent of a studio, network, or channel.

Shortly after *The Walking Dead* attracted record-setting audiences, HBO at last equaled—and by some measures, surpassed—the thirteen-year-old benchmark set by *The Sopranos*. *Game of Thrones* (HBO, 2011–) became HBO's most-watched series in the United States—attracting millions of viewers and continuing to set ratings' records for new episodes as this book went to press in the series penultimate season.[1] *Game of Thrones* also earned the undesirable distinction of being the most pirated show around the globe. In response to fans' demands, and seeking to stem unauthorized access, HBO made the series simultaneously available worldwide beginning in its fifth season. Such a case suggested the possibility of a new type of television series—the global blockbuster.[2]

Based on the transnational reach internet distribution enables—most evident in the global strategies of Netflix and Amazon Video—we might expect internet distribution to be a key cause of *Game of Thrones*' international popularity. But this is not the case. HBO developed the series into a global blockbuster using traditional television distribution relationships with cable channels and broadcasters around the world.

The internet is important to this story because it enabled a level of piracy difficult in a pre-internet era and for the way its social media functions served as kindling that fueled unauthorized access among fans. The fervent demand by viewers everywhere to see episodes simultaneously led the series to become the first to command worldwide, concurrent attention.

This early in the development of internet-distributed video, no legacy company other than HBO could have responded to the passionate global interest in a series to accomplish simultaneous airing in 170 markets. HBO was both the series' producer and first market distributor, which allowed it to adjust the timing of *Game of Thrones*' international availability without the tension that would arise if an external studio had produced it.[3] HBO also had significant international reach and relationships that provided it with a direct pipeline to viewers outside the United States.

Although HBO was at the forefront among the legacy television industry in experimenting with internet distribution through its HBO Go (2010) and later HBO Now (2015) services, it isn't clear that simultaneous worldwide airing of *Game of Thrones* was part of its plan. The simulcasting didn't begin until the premiere of the fifth season. Reducing piracy was likely the greatest motivation, although it maintained its status as most pirated even once simultaneously available. Consistent reporting of the show's high piracy did provide valuable publicity for the series and affirmed the reputation of the service, but HBO also certainly desired to convert those watching unauthorized streams into paying subscribers and to maintain good relationships with those channels that licensed the show in other countries. Its willingness to allow non-HBO-owned channels to simulcast earned it goodwill by providing a tool that might coax potential subscribers/viewers to become part of the authorized viewing audience.[4]

Rampant piracy of a budget-busting blockbuster hit was unsurprising, but the emergence of social media as a venue for shared viewing and communication that mandated simultaneous global availability was a less-anticipated shift in the broader environment of television viewing. Of course, many accessing unauthorized streams of the show certainly did so to avoid paying for the service. Fans in time-delayed markets, however, were also motivated because they could not avoid spoilers in their social media feeds.[5] Many series with devoted fans had maintained web-based discussion groups for years that were made up of fans accessing new episodes at different times, but the ease of unauthorized streaming of rich video files was a recent development. The scale of social media use and the tendency to watch with a mobile device in hand arrived to exacerbate the ease of unauthorized access.

Game of Thrones was admittedly not the first to aspire to worldwide simulcast of a scripted drama. *Doctor Who* attempted a similar feat for its fiftieth anniversary, simulcasting an episode in November 2013 in seventy-five countries. But unlike the special-event occasion of *Doctor Who, Game of Thrones* suggested a possible new model of international distribution. Simultaneous release significantly disrupted decades of established practices for international television distribution. It seemed unlikely it would become the norm, but it was certainly now a possibility. HBO's strong brand, its ownership of several channels in other markets, and its practice of producing—and thus owning—its series, positioned it well to begin exploring new digital and global media practices.

Game of Thrones also possessed a variety of features likely to fuel global popularity and desire for immediate viewing. The series debuted at a time when excellent drama circulated extensively around the globe. The distinctiveness achieved by many original U.S. cable series made U.S. product more than just something to fill out linear schedules in other countries. The historically insular U.S. market was even watching Danish, French, and British drama, although more often buying successful drama formats and remaking them.

Among this international surge of sophisticated drama, *Game of Thrones* "traveled well"—that is, its fantasy setting was not geographically or culturally distinctive, freeing it from seeming particular to the U.S. market. Traditionally, such epic and noncountry-specific tales had been among the most successful series and made-for-television films in international trade.[6] HBO's budgets enabled the lavish spending that made *Game of Thrones* visually breathtaking to an extent that was less feasible for ad-supported services. The series also drew from a known and acclaimed series of novels that helped drive initial interest in the series.

Certainly the on-demand access enabled by internet-distributed HBO Go should be acknowledged, but it was not critical to *Game of Thrones'* uncommon success. HBO Go launched in February 2010 at no additional cost to U.S. subscribers, and *Game of Thrones* was the first major series HBO debuted after launching this cautious test of internet distribution. HBO was immediately able to offer an extensive library of series and films produced by HBO for HBO, as well as its current series, and it was able to license much of its theatrical fare. In launching HBO Go, HBO followed the same strategy that had motivated its move into multiplexing and then VOD in previous decades. As a subscriber-funded service, it sought to add value by making it easier for subscribers to watch what they wanted when they wanted—not simply what it chose to schedule at a given hour.[7] HBO Go made *Game of Thrones* easier for U.S. viewers to watch however and wherever they desired, but it was limited to U.S. IP addresses. At the end of 2012, HBO's parent company, Time Warner, launched a venture with HBO Nordic that offered its first non-U.S. internet-distributed service—much like what would become HBO Now in the United States.

Game of Thrones relied on global interest and HBO's extensive international holdings to suggest the possibility of television series as global blockbusters. But even before its simulcast, Netflix had begun shifting its business to suggest similar aspirations. Netflix first surprised the industry with its early and aggressive move into producing original series. That first U.S. series—*House of Cards* (Netflix, 2013–)—accomplished Netflix's aim of solidifying the legitimacy of the service in the United States. But the pivot into original production with *House of Cards* also provided an early hint that Netflix too had its eye on global distribution. It hadn't yet created a global blockbuster, but the international reach its service achieved in early 2016 positioned it to become the first global television "network."

The buzz about *House of Cards* emerged nearly two years before its episodes. Rumors that Netflix had outbid HBO and AMC and committed a jaw-dropping $100 million for two seasons of a series directed by David Fincher and starring Kevin Spacey first appeared in blogs in March 2011. This was the high point of Netflix's surge into the popular consciousness, as it grew its U.S. subscribers at a rapid pace, and still months before its major stumble in attempting to split the by-mail and streaming service. The rumors were confirmed within three days and the truth was as momentous as preliminary accounts suggested.

The news did exactly what Netflix sought by drawing even more attention to the service. Of course much of that attention was nay-saying skepticism, but even skeptical buzz was helpful for a service trying to find its way into American homes. This announcement was important because it revealed Netflix's aspirations to be an originator of content, but it was earth shattering because the purported budget and two-season commitment defied the norms of series creation. The first season was scripted and developed by Beau Willimon in coordination with a studio known for independent film production—Media Rights Capital. It was virtually unheard of for a series to be accepted for a full season without first screening a pilot—and certainly never with a guaranteed second season.[8] With such an offer, Netflix raised the stakes of what top talent could ask for across the industry.

Although *Game of Thrones* would match and surpass this budget when it debuted a month after this announcement, the roughly $3.8 million per-episode budget likewise made clear that Netflix was not going to offer cut-rate series. Unlike cable channels' first forays into original series production a decade earlier—in which they attempted to produce cheaper versions of broadcast network-style shows—Netflix immediately targeted distinction. *Game of Thrones'* cost owed considerably to the expansive cast, special effects, and location shooting, but *House of Cards'* budget bought it a cast and creative team that would generate attention. Acclaimed film talents Fincher and Spacey added to the intrigue. The prospect of *House of Cards* signaled Netflix as a television competitor, yet it did so in a way that kept it tethered to its still-stronger identity as a film distributor.

In the first interviews confirming that Netflix had ordered two seasons of *House of Cards*, the service suggested that it was also likely to deviate from conventional, one-episode-per-week delivery. At the time, chief content officer Ted Sarandos suggested Netflix might make four or more episodes available at once. When it ultimately debuted, Netflix released the full season simultaneously. This kicked off another round of analyst speculation as to whether this was strategically wise and what this would mean for the future of television. Few acknowledged that Netflix was building a library, which would require different strategies than those involved in creating a schedule, or that the measures of success were different for Netflix because of its subscriber-funded revenue model and on-demand distribution capability.

Gauging the actual success of *House of Cards* is difficult, precisely because Netflix's internet distribution and subscriber-funding made it different from other television competitors. The buzz created by the show's announcement and eventual launch provided a significant amount of promotion for the series. And releasing all the episodes at once led journalists to write about the emerging phenomenon of *bingeing*: consuming all the episodes in a matter of days or even hours.

Where a sense of the cultural significance of *Game of Thrones* and *The Walking Dead* was evident in both ratings and piracy data, the scope of

House of Cards' influence was more difficult to determine. Like many other notable shows at the time, it was clear that *House of Cards* was necessary viewing among a subpopulation, or what Netflix would come to call a *taste community*. Appraising the scope of that population and how it contributed to adding or maintaining subscribers was difficult to estimate without access to Netflix's data.

But *House of Cards* also didn't need to inspire immediate viewing to be valuable to Netflix. The series would remain an asset available in Netflix's library for some time. The series was just as valuable to Netflix as part of an accumulating push to subscribe even if it didn't drive immediate subscription or viewing. *House of Cards* also revealed some of the conundrums of evaluating the success of internet-distributed television: Did it matter how many viewed it in the days, weeks, or months immediately following its release? How do we assess the value of a show that becomes part of a library in perpetuity?

House of Cards illustrated that original series competition from the internet-distributed services was already becoming a reality in 2013. An ample budget and well-crafted creative vision could lead to top-rate television regardless of the service distributing it. That Netflix, Hulu, Amazon Video, and others were part of the television industry was not yet obvious. Their internet distribution still led many to believe they were more different from television than they actually were.

For all the ways Netflix defied industry norms with *House of Cards*, it is also important to acknowledge the conventions it did maintain. Netflix paid a steep licensing fee for *House of Cards* to maintain the U.S. distribution rights for longer than usual, but it did not secure the international or DVD rights. These aspects reveal *House of Cards'* status as an early offering from Netflix—one primarily about expanding its U.S. reputation and subscriber base. By the time Netflix offered later series, such as *Marco Polo* in 2014, its global aspirations were more apparent; it contracted for global, long-term rights in a way that would make Netflix the only legal source for the series, rather than merely its first window.

The fate of Netflix's *Marco Polo* underscores the limits of focusing on the exceptional—series such as *Game of Thrones* and *House of Cards*. By many measures, *Marco Polo* could have—even should have—been as revolutionary as *Game of Thrones* was for HBO. By its 2014 release, Netflix was available in forty countries. This provided strong global distribution reach, especially for a company that had not even existed twenty years earlier and had been a player in the television industry even more briefly. *Marco Polo* offered an epic and widely accessible tale. Netflix's subscribers outside the United States were growing faster than those in the United States by 2014, so such universal reach seemed a good strategy.[9] Yet even with a budget of $90 million for ten episodes, the series failed to generate much buzz. The series' quiet existence revealed the uncommon accomplishment of HBO's trifecta of compelling content, distribution flexibility, and global reach.

HBO's aggressive embrace of changing competitive conditions was responsible for *Game of Thrones* becoming the first global television blockbuster. *House of Cards* is more relevant as an "important first" that illustrated how internet-distributed services could produce content comparable to top "television" fare. Without a doubt, *House of Cards* was symbolically very important as Netflix anted up and bet the house in announcing its entrée into scripted original series. But Netflix's guarded approach to releasing viewership data makes it impossible to assess how significant *House of Cards* was in the steady uptick of subscribers that followed the company's expansion into original series production.

The arrival of *House of Cards* also requires a shift in terminology, from referring to networks and channels to discussing all these entities as content services. Rather than the cumbersome delineation of HBO as a channel and Netflix as a service, all these providers become more manageably considered as program services. This is especially the case as HBO Go and HBO Now untether HBO from scheduled distribution. The term *portal* distinguishes internet-distributed services. Just as stations and networks organized broadcast programs and channels organize cable, portals organize internet-distributed television.[10]

Game of Thrones and *House of Cards* offer a slightly different insight into emerging competitive norms, but there are many commonalities that unite them as well. Neither was as much of a long shot as other series we have discussed. Both were privileged to emerge from services that had the reputation and the financial support to give them better-than-even odds. Despite all this, the scale of success these shows accomplished was not predicted.

These series revealed the business practices shifting behind the scenes as services recognized the importance of self-producing their series and designed shows for a simultaneous international audience. Unlike the dominant model in television in which hits were built on selling series again and again around the world and over time, these series were intended to produce exceptional value only for the services that originated them.

Beyond the behind-the-scenes changes in how television studios sought to earn profits, these series also revealed a shift in the logic of what studios were trying to accomplish in making shows. FX president John Landgraf described the transition succinctly in 2015 when he explained that the organization of television in a schedule had made it somewhat disposable. As a result, creatives and executives thought about their mission as trying to make a show that audiences would watch on Sunday night. But television in an environment of internet distribution had to set its sights on a different aim. Studios needed to make not simply television shows people would watch on Sunday night, but television shows people would watch and talk about ten years later.[11] Such an endeavor didn't suggest a change from the pursuit of "distinction," but it explained how and why series were changing.

26 TV Goes Indie?

At the other end of the spectrum from global blockbusters is what might be called indie—short for independent—television. The scarcity of broadcast and cable technologies limited the availability of television to an extent that independent television was unusual in the United States—the closest example was found on local cable access channels. But internet-distributed television had much greater capacity, and services quickly developed that made sharing video with a sizable and dispersed audience feasible.

YouTube became the primary distributor of what started simply as amateur content and gradually developed into a semi-professional and even professional sector. There had always been more stories and storytellers than could fit on television schedules. Open access distributors such as YouTube provided a way to share those stories.

Some people predicted that YouTube would destroy the television industry in the early years of the twenty-first century. This reflected the view that new media would kill off old media. Although billions of people came to view video on YouTube, this viewership did not overthrow the legacy television industry in the way some expected.

YouTube has been so many things—often simultaneously—since its launch that it is difficult to speak of it coherently. YouTube initially proved valuable to the legacy industry as a way to distribute clips and other bits of content that promoted its shows. YouTube also became the source of amateur video that could develop audiences that rivaled legacy television sources. Some amateurs merely used the site for distribution—as a way to share—while others distributed videos through YouTube with a goal of financial profit. While pursuit of commercial ends is often the distinction between amateur and professional behavior, a significant sector emerged on YouTube that was not amateur, but was also not professional by the conventions of the legacy industry.[1]

YouTube is different from the other portals discussed here because it functions as a reasonably open access distributor of video. It does not develop and intentionally curate programming in the way legacy television providers and portals such as Netflix, Hulu, and Amazon Video do.[2] It also relies on a different business model. YouTube neither pays for development nor owns a stake in the content it distributes. This provides far lower operating costs and enables YouTube to support itself through an advertiser-supported revenue model. YouTube's scale as the dominant open access distributor also makes relying solely on advertiser support feasible.

YouTube was not alone; other services that were built on open distribution also emerged. By the end of 2016, the most profitable open access content built a following around a personality—PewDiePie, Bethany Mota, and Hank and John Green, to name but a few. Such personalities distributed programming that mostly provided a variation on the legacy television forms of the talk show and reality television amid the growing scope of internet-distributed television. Top personalities earned significant advertising and sponsorship revenue—enough to make video creation a professional and vocational endeavor.

YouTube and other open access portals could also be used to distribute "independent" scripted series. The higher costs of scripted series production made recouping budgets through advertising difficult and made these series less viable from a business standpoint despite important artistic and creative accomplishments. As of 2016, open access portals' most substantial impact on the business of scripted television was as a route into the legacy industry. Creatives such as Issa Rae (*Misadventures of an Awkward Black Girl*, *Insecure*) and Lucas Cruikshank (*Fred*) leveraged storytelling personalities distributed first on YouTube to launch series or films on HBO and Nickelodeon, respectively. In other cases (e.g., *High Maintenance*, *Broad City*), series developed by independent industry outsiders moved their projects to legacy distributors.

By creating an opportunity for distribution without legacy media's gatekeepers, YouTube and similar sites enabled an indie television sector that had not previously been feasible because of the rigid control broadcasters and channels held over the tightly constrained schedule. Once YouTube created an open access platform, the primary challenge shifted from distribution to discovery. Just as the deluge of content among legacy services drowned even the most distinctive series, the vast ocean of YouTube content made it difficult for innovative content to find an audience.

As a result, the cases of indie television that had the greatest impact on the business of television came from creatives who were already established entertainment figures. Joss Whedon's *Dr. Horrible's Sing-Along Blog* offered a case of short-form video as early as 2008. Over seven years later, another established creative used the opportunity of internet distribution to create an independent series of similar aspiration. It arrived unexpectedly on January 30, 2016, as an update titled "A Brand New Thing from Louis C.K." on his website and pushed by email to newsletter subscribers. The message channeled Hemmingway:

> Hi there.
> *Horace and Pete* episode one is available for download. $5.
> Go here to watch it.
> We hope you like it.
> Regards,
> Louis.

The message offered no suggestion of what *Horace and Pete* was. C.K. had a solid fan following based on his stand-up and his FX series *Louie*. He had experimented with direct distribution in December 2011 when he released his "Live at the Beacon Theater" stand-up special through his website for five dollars. The *Horace and Pete* experiment quickly drew attention from television critics, although minimal general publicity.

Horace and Pete tells the story of two brothers (later revealed to be cousins) who run a small New York neighborhood bar that has been in the family for many generations. The series was arguably the first case of a creative with considerable professional and celebrity clout adopting a direct-to-consumer strategy with scripted programming. The endeavor was wildly disruptive. Although services such as Netflix significantly altered familiar distribution practices of broadcast and cable television, C.K.'s experiment eliminated even the studio's role. This challenged production norms much more extensively than digital distribution. Both unprecedented and unanticipated, it was difficult to know how to evaluate the experiment.

Several journalists writing about *Horace and Pete* referenced Beyoncé's self-titled iTunes album release in December 2013, which had similarly simply appeared without any of the typical prerelease promotion fanfare. The album achieved 1.3 million sales in seventeen days and was quickly heralded as a massive success.[3] The comparison was apt in terms of a shocking lack of advance promotion, but otherwise Beyoncé's album was a studio product released on the dominant distributor of digital music. In contrast, *Horace and Pete* had all the attributes of a professionally produced television series, but it had been created without any of the production or distribution infrastructure.

Just as abruptly as it arrived, *Horace and Pete* posted a tenth episode, C.K. announced that this was a full season, and soon after explained there would not be further episodes. Most of the media coverage claimed that C.K. was now millions of dollars in debt. As part of a belated promotion blitz, C.K. explained on *The Howard Stern Show* that he had intended to risk $2 million of his money to fund the first four episodes and expected that he would have earned enough from the streaming of the early episodes to produce the remaining ones.[4] Revenue was not as swift as desired, and he had to take out a loan to finish production.

Unlike the case with the "Live at Beacon Theater" release, in which he posted information about the speedy pace of download sales, C.K. offered no information about the viewership of the series. He revealed that the series cost $500,000 per episode (less than a sixth of a typical hour of scripted television), meaning that he produced ten episodes for a rough cost of $5 million. Several other features of the production were unclear: Did that $500,000 cost include any remuneration for C.K., who served as producer, director, writer, and actor? Was the production team paid at union rates? What about the fees for top talent such as Alan Alda, Edie Falco, and Steve Buscemi?

Gradually C.K. offered more insight into his motivation for the experiment. He avoided promotion because he was curious about whether a show could be made available without the audience being clued in to what they would experience. He wanted the audience to encounter all aspects of the series as a surprise and without expectation.

An experiment such as *Horace and Pete* must be assessed by many measures. It was arguably an artistic success. Critics uniformly praised the series and it won a Peabody Award. As producer, director, writer, actor, distributor, and financier, C.K. alone called the shots to a degree unimaginable in U.S. television. *Horace and Pete* was a commercial success. C.K. recouped his investment within the year and acknowledged sharing revenue with the other actors who held a stake in the show.[5]

Horace and Pete was also successful as an experiment illustrating the creative boundaries that alternative production practices could push. Somehow, the episodes were simultaneously television and yet far removed from what viewers typically saw on screen. The series harkened back to television's earlier days. *Horace and Pete* had a decidedly live feel. The pacing was far more theatrical than contemporary television. The script relied on long speeches—practically monologues—that brought radio drama to mind. There was far more talking than action in *Horace and Pete*, which wasn't really "about" anything more than family relationships. The culmination of these factors produced a series that was uncommonly, even hauntingly intimate in a way that made it very different from nearly all other television, despite the fact that intimacy has long been understood as a central characteristic of television.

Beyond these measures of evaluation, *Horace and Pete* was important for its defiance of convention at the peak of the television ecosystem. Internet distribution had made the production of independent television series possible, but few ambitious experiments with known talent emerged. With *Horace and Pete*, C.K. suggested a possible alternative to industrial norms, opened the minds of audiences and creatives to self-distribution as a possibility, and made it infinitely easier for someone to produce and distribute the next such project.[6]

To be a success *Horace and Pete* didn't need to launch a new a trend of self-production and distribution. C.K. regarded it as a success as a creative exercise, but it was much more than this. It is easy to imagine that the common sense of the industry would have believed a project like *Horace and Pete* impossible. But now a test case existed that illustrated both the opportunities and limitations.

27 The End of the Early Days

Netflix, video on demand, and over the top were the big developments that initiated changing strategies for transmitting television to viewers during the first years of internet-distributed television. These shifts in how viewers could watch were joined by the emergence of global blockbusters, indie series, and a wide assortment of internet-distributed original programming to expand what viewers could watch. The legacy industry made quieter shifts in revenue models and business strategies in these years as well. Although less noticed, these were designed to evolve existing businesses and pivot into new ones.

The legacy industry adjusted its businesses through expanding nonadvertiser-based revenue by establishing self-owned studios and buying cable channels outside the United States. The balance between revenue from subscribers (in the form of carriage fees paid by service providers) and advertising gradually tipped toward relying on subscribers. FX president John Landgraf noted that when he began at FX in 2004, 55 percent of the channel's revenue came from advertising. By 2015, advertising accounted for just 34 percent and was diminishing by a percentage point a year.[1]

In that same period, the broadcast networks transitioned from being entirely ad-supported to relying increasingly on *retransmission fees*, which were essentially the same as the subscriber fees paid to cable channels (see chapter 18). Retransmission fees accounted for 15 percent of broadcast revenues by 2013, and were projected to grow to 25 percent by 2019.[2] With little notice, ad-supported cable channels and broadcast networks had become more accurately described as subscriber-supported and advertiser-subsidized, although they maintained and even expanded the advertising load of their schedule.

Legacy television companies were not complacently waiting for internet distribution to have the devastating effect on their business that the recording and newspaper industries had experienced. The shifts in revenue streams weren't readily obvious, so the extent to which they had begun building business models that were different from their previous norms was rarely acknowledged.

The slow pace of change led some to believe the industry was in denial. But the situation was complicated. It wasn't in the financial interest of legacy companies to move quickly into internet distribution. These companies were still making a lot of money in their legacy businesses and sought to maintain those revenues as long as possible. The trick was to be prepared to offer internet-distributed portals as soon as substantial consumer demand emerged. Many of these companies spent extensively on initiatives and services in the earliest days of internet video—2005 to 2007—that were quickly forgotten and went unused because they arrived too early. The pace of adjustment evident by the time the tipping point arrived in 2015 suggested that most companies were not idly passing these days but were discretely preparing for an internet-distributed future.

A comparison with the passenger rail industry was often made during these years of changing television distribution. This recalled the failure of the passenger railroad business to act when the airline industry emerged. Seeing itself as only in the train business—rather than in the broader human transportation business—those in the passenger rail industry didn't perceive human transportation by plane as a threat, and the airlines quickly decimated the long-haul passenger train industry. Provocateurs likewise challenged the television industry—particularly the broadcast networks and cable channels—to recognize that they were no longer just in the television industry: they were now also video content providers. Remaining committed to schedule-based television risked losing an audience to those offering an on-demand experience.

From outside the giant content corporations, it was difficult to know how extensively companies were preparing to pivot their distribution once internet distribution became a broader phenomenon. Their preparations were made quietly—which only served to feed the sky-is-falling attitude perpetuated by the analysts. Public statements by content-owning industry leaders remained bullish on legacy practices until the moment extraordinary pivots were announced.

Despite much greater sophistication in understanding what was and wasn't happening as a result of internet distribution of video, it still was not uncommon to see headlines such as "The Internet Is Slowly but Surely Killing TV" in 2015 as death match narratives persisted.[3] Sometimes the articles actually offered more nuanced assessment. Those who simply scanned headlines could not be blamed for assuming the end of television was yet again or still upon them.

Beyond the natural curve of technology adoption, access to sports programming provided the primary bulwark against change. Services such as Netflix, Hulu, and Amazon Video could license and distribute scripted series and movies, but sports viewers could watch beloved contests only on cable and on broadcast channels that had paid multiyear licensing deals. The cable channel ESPN in particular stemmed the tide against cord cutting with its exclusive access to several professional leagues' games. ESPN was consequently able to command significant subscriber fees of as much as $6 or $7 per subscriber per month, even for homes that weren't interested in ESPN's programming, because of the way cable was packaged.

Frustration with cable packages intensified as the new technology of internet distribution seemed to provide an alternative. Calls for "unbundling" packages grew louder. Cable companies threatened that unbundling would force prices for individual channels much higher because viewers that didn't want the channels in the first place would no longer be forced to subsidize them. Instead of the $6 per month per subscriber paid to ESPN, the cost might rise to $30 a month if those who didn't want the service could opt out. Would viewers pay $30 a month for ESPN? Maybe. Or maybe ESPN would change what it was willing to pay for sports league licenses. Or maybe sports leagues would cut out the middleman of ESPN and launch a portal that reached viewers directly as did World Wrestling Entertainment in February 2014. The fallacy of saying that ESPN would cost $30 a month without the bundle was that it didn't allow for a major reconfiguration in television's business model. It was becoming clear that such a reconfiguration was an obvious—if painful—necessity.

Netflix was largely responsible for this change, but as calls to "unbundle" television grew, Netflix itself was no different. On one hand, Netflix disrupted the decades-old experience of television and offered a value proposition that heightened viewers' barely contained dissatisfaction with multichannel pricing and packaging. On the other, Netflix too aggregated a bundle of content. Just as HBO built a successful subscription business from creating content for clusters of viewers with tastes underserved by other television services, Netflix used a "conglomerated niche" strategy of providing a little bit of content for a lot of different audience segments.[4] It primarily distinguished itself with some content otherwise unavailable (Netflix's original series). It also offered a distinctive viewing experience at a relatively low price point.

Subscriber embrace of the Netflix bundle amid steady complaints about multichannel bundles superficially seemed paradoxical. On closer consideration, it revealed that viewer discontent with old models was not about bundles so much as the value proposition of the bundles offered by multichannel providers. Netflix also benefited from the fact that few consumers considered the monthly fees they paid for internet service as part of the cost of Netflix.

Netflix's profit margins weren't large, a point regularly hammered on in comparisons to HBO ($.28 versus $3.65 per subscriber per month in 2014).[5] Like e-retailer Amazon, it staked its early strategy on mass reach even if it had to keep profit margins low to infiltrate homes on a massive scale. This was a long-range strategy. Once the sector matured, Netflix's market dominance would make it difficult for new competitors to enter. It could then raise fees with little consequence.

Netflix's value proposition was also dependent on its library of content. The company boldly entered the legacy business in 2013 with original series and proved it could compete. Such moves pressured the early détente. There was no sign that the pace of change would soon accelerate and that a tipping point was near as 2014 ticked to a close. Then without warning, the legacy industry revealed its preparation to join the ranks of internet distributors, and the détente was over.

28 Portals: The Beginning of the Middle Days

By 2015 it was clear that internet-distributed television was crystalizing as its own sector of the television industry. Internet distribution affected various segments of legacy television differently. It was great for viewing scripted series, but had little impact on television programming such as news and sports or television just kept on in the background. The expansion in internet-distributed services consequently challenged only some sectors of the television industry.

In 2015, it became clear that the legacy industry did not have its head in the sand. It was preparing—or even was already prepared—to pivot into internet distribution. The industry was on the precipice of change. Maintaining the status quo rewarded its entrenched interests, so the legacy industry wasn't leading the way. The contours of the future of television were becoming visible, but the pace of transition remained unclear.

Little new happened in 2015 that explains why this year became an inflection point. One notable development was some resolution of questions about what regulatory norms would govern internet distribution. Uncertainty about *net neutrality*—a policy that would prevent discriminating against different users by limiting, blocking, or slowing down service, or imposing discriminatory pricing—hampered innovation because of its potential effects for business models. Discussion of net neutrality dated to the early 2000s and was initially related to the fear that internet distributors might discriminate access to content based on particular political ideas. But by 2010, the bigger threat was that internet service providers would develop tiered service speeds that would require internet-distributed services to pay fees for quick and efficient access to their sites and services.

Without a neutrality policy, providers could charge sites to make sure their content loaded quickly—what was being called an internet fast lane.

The imposition of such fees would have significant consequences for services using the internet to distribute television by adding to their operating costs to an extent that would require profound adjustment in business strategy and possibly raising prices. Internet-service providers longingly eyed this additional revenue stream.

The FCC moved slowly in establishing the playing field for internet service. After the courts threw out early efforts at light regulation written during the Bush administration, the FCC of the Obama administration attempted rulemaking that would withstand judicial review in 2014. The policy announced in early 2015 upheld net neutrality by classifying internet service in a way that forbids providers from discriminating among services—which is what would happen if sites were able to pay for faster access. More important, it at last seemed to establish the playing field for competition. The rules weren't the strongest policy option—a legislative act by Congress would enshrine the regulation more certainly. Within months of the Trump administration taking office, the Trump FCC announced plans to eliminate net neutrality policies. Neutrality policy was not the regulatory outcome the internet service providers desired, but eliminating the policy uncertainty invigorated innovation among content providers, at least for a few years.

Change also resulted because of the disruption of the initial symbiosis between the legacy industry and new internet-distributed services. With its launch of both *House of Cards* and *Orange Is the New Black* in 2013, Netflix proved itself a formidable content creator out of the gate. Amazon Video also launched original series in 2013, although to much less fanfare. If internet-distributed services were to create content, then the content creators needed to more aggressively launch their own internet-distributed services.

Surprisingly, it wasn't new entrants into the television industry that knocked over the first domino. The closest it came to an inflection point was when legacy entities CBS and HBO announced in late 2014 that both companies would bring internet-distributed portals of their programming to market in 2015. Their announcements were unexpected and made within days of each other, which escalated the sense that this was the beginning of significant change. These services suggested how legacy television could evolve—although HBO and CBS couldn't have been more different as content services.

HBO's development of a standalone internet-distributed service was long overdue. Many viewers begged the company to allow viewers to subscribe directly to just HBO Go when HBO introduced its internet-distributed VOD supplement in 2010. HBO described its motivation as creating a service for those uninterested in a traditional cable package. Research commissioned by HBO found two groups who didn't subscribe to HBO but could be persuaded to do so: the roughly ten million homes that would subscribe if there was a cheaper way, and the ten million internet-only homes.[1] HBO Now served both groups.

HBO may also have been motivated by a desire to transition traditional subscribers to HBO Now as the service had the potential to allow HBO a greater percentage of subscriber fees. Few subscribers realize that half of the fee they pay for HBO typically stays with the multichannel provider. HBO Now enabled subscribers to access the service without dealing with a cable provider, leaving HBO with a greater share of the subscriber fee.[2]

HBO Now launched in April 2015, had 800,000 subscribers ten months later, and had two million by the end of 2016.[3] Showtime launched its similar Showtime Anytime service three months later as subscriber-supported services once again led innovation in making content more convenient. The slow growth in subscribers was unsurprising and was not a sign of failure for those still distributed by multichannel services as well. A key aim of the portals was to make their content available to homes eliminating cable service, and this remained a small, though increasingly significant, percentage of households. The internet-distributed version of the service might dominate in time, but that wasn't expected to be the case in the near term.

CBS's announcement that it would create CBS All Access as a service for distributing current CBS series and the conglomerate's decades-old catalog of series was the more surprising of the announcements, and all the more significant because CBS seemed the least likely to introduce an internet-distributed service. CBS was the dominant broadcast network throughout the 2000s and achieved that status with programs nearly the opposite of the distinctiveness found on cable. CBS wasn't a network associated with cord-never millennials, so the announcement of CBS All Access was a bit bewildering.

Admittedly, CBS had been at the forefront of internet distribution with its NCAA men's basketball March Madness Tournament coverage and had been well rewarded with record viewership of the regular broadcast of the games in addition to the expanded audience reached on a multitude of devices through internet distribution. Viewing of events like the NCAA basketball tournament and the Olympics was limited by broadcast and cable technologies that could transmit only one event or game. Internet-distribution enabled the distribution of more competitions, and fans flocked to them.

CBS All Access slowly rolled out on different devices throughout 2015. Initially it was a $6.99-per-month service that still had advertising. It included both an on-demand component available everywhere and the ability to stream live CBS programming for viewers living in cities served by stations owned and operated by CBS.[4] CBS needed to establish agreements with its affiliates to allow streaming in other cities because the streaming service risked taking viewers out of the broadcast audience, which would diminish local stations' ad-revenue. New programs were available on demand to all subscribers the day after they aired. Notably, the National Football League games CBS licensed were not available through All Access because CBS didn't have internet rights to the games.

There was little evidence that All Access was widely popular; six months after launch, CEO Les Moonves deadpanned that he'd reveal how many people subscribe as soon as Netflix released how many people watch its shows. A little over a year after launch, analysts estimated the service had 500,000 subscribers, growing to nearly 1.5 million a year later, in February 2017, when CBS began making high-profile content available exclusively on the service.[5]

As with HBO Now, the subscriber base was small but important because these viewers likely had no access to CBS otherwise. Moreover, All Access began to provide CBS with information about what people watch and how they watch that was more granular than what Nielsen offered—although it certainly was not broadly representative.

CBS marketed All Access as a way to watch CBS's current programming, but it was more interesting as an experiment of a studio-owned portal. Unlike Netflix, which had built its business as a distributor of library content licensed from other studios, then gradually developed its reputation as an exclusive source for a few original shows, CBS possessed a vast library of shows produced by CBS Television Studios, Paramount, and studio libraries purchased as the conglomerate grew. CBS All Access illustrated how a studio might develop its own internet-distribution business based on self-owned content instead of licensing content to Netflix.

The other broadcast networks had tested the waters of internet distribution through the joint enterprise of Hulu, which significantly predated these 2015 developments. Hulu's complicated origin as a joint enterprise of Fox, NBC, and ABC had long made Hulu's strategy in the emerging landscape unclear. The service arguably found its footing in 2015 when it shifted to subscriber funding and produced original series that gained some notice. Hulu had twelve million U.S. subscribers by May 2016, up from just five million at the end of 2013.[6]

Hulu was one of the first internet services available. Any joint venture is tricky, but its attempt to bring three of the major television groups together before anyone had much sense of how internet distribution would change the television business was particularly challenging.[7] Where Netflix and Amazon Video could scale their businesses by amassing a global audience, Hulu's owners—companies that were earning billions selling shows to channels around the world—were more wary of disrupting the established pattern for selling series in international markets. Just as CBS All Access was a U.S.-only portal, making Fox, NBC, and ABC-owned series available on an internet service outside the United States would diminish the licensing deals international networks paid for their series and violate existing agreements.

Table 28.1

Top ten internet-distributed subscriber services (October 2016)

Service	Launch	U.S. Subscribers
Netflix	2007	44.74M
Amazon Video	2006	76.2M*
MLB.TV	2002	3.5M (end of 2015)
WWE Network	2014	1.01M (373K non-U.S.)
HBO Now	2015	2M
Crunchyroll	2007	1M (worldwide)
Showtime	2015	1.5M
CBS All Access	2015	1.2M

Source: Parks Associate OTT Video Market Tracker

*Amazon Prime members, not necessarily Video users

Beyond the general programming offered by HBO, CBS, Hulu, and Netflix, more specialized portals also multiplied. By the end of 2015, there were nearly one hundred portals available in the United States.[8] World Wrestling Entertainment launched the $9.99 per month WWE Network in 2014 that was available in 140 countries, and Viacom offered Noggin—an ad-free service with hundreds of episodes of programs for preschoolers for $5.99 per month in March 2015. Comcast-NBCUniversal launched SeeSo, an ad-free comedy portal for $3.99 a month that offered new content from well-known comedy talents and access to NBC-owned comedies such as *Saturday Night Live* and *Parks and Recreation* in January 2016. Disney introduced the DisneyLife portal at the end of 2015, which offered Disney movies and Disney Channel shows to viewers in Europe and China.[9]

Only the handful of portals that offered broad libraries—Netflix, Amazon Video, Hulu, HBO Now, maybe CBS All Access—were well known. Many others, such as Trout TV, Hungry Monk Yoga, and Chris Cox Horsemanship, offered hyper-niche content. Both the broad and narrowly targeted portals could be profitable; those with more basic programming also had far lower programming costs. The portals were distinct from channels and networks because they were internet-distributed, but also because many—seventy-six of those on the market in 2015—were purely subscriber funded. This previously uncommon revenue model became surprisingly abundant and arguably distinguished these portals just as much as their distribution technology.

Viewers were motivated to try internet-distributed services for an array of reasons. They provided a viewing experience with not only greater convenience and control, but also distinctive content. Many appreciated the convenience of being able to turn tablet and mobile phone screens into televisions, and they often featured much better search functionality than multichannel provider services. By August 2015, half of U.S. homes had a device that connected their television to the internet, so the capability to receive television by internet had reached a tipping point for living room screen viewing as well.[10]

By 2015 it was clear that innovation in television—whether from newcomers or legacy entities—would be focused on internet distribution. These were the heady, Wild West days of an emergent industry. The flurry of portals introduced in 2015 and 2016 indicated the legacy industry's recognition of where the business was going. They didn't seek to hasten viewers' adoption of internet-distributed television, but the number of homes without cable had become too substantial to ignore. The legacy industry, which had been built on selling series again and again, avoided internet distribution of highly valued series as long as it could. By 2015, the risk of being inaccessible—especially in an environment of a surplus of programming—became too great for it to remain on the sidelines.

Portals provided a valuable opportunity for the studios that were mostly ensconced in large media conglomerates. In the extreme, it allowed them to transact directly with their audience—although most were not yet willing to develop the customer-service infrastructure to do so. Portals quickly became the solution to internet distribution for those who owned content. But other companies also sought to stake a claim in the shifting distribution environment.

29 The Unbundling Continues

The change that exploded in 2015 was even bigger than the arrival of internet-distributed portals. At the same time that portals enabled new ways to access content, other companies believed there was a market for internet-distributed packages of channels. The first service to target the cable bundle entered the market in February 2015. Sling TV initially offered twelve core cable channels for only $20 per month, but it doubled the available channels by mid-year and offered subscribers additional packages of themed channels (kids, films, sports) for an additional $5.[1] Sling TV was available on a variety of screens and out-of-home devices, but it had no equivalent of a DVR. Most of its channels featured the same on-demand access that would be available from a cable subscription.

Sling TV provided the first of what came to be called *skinny bundles* or *virtual MVPDs*.[a] Similar to conventional cable offerings, such services offered access to a package of channels, but distributed the channels over the internet. By offering fewer channels for a lower cost, they aimed to bring competition to the largely uncompetitive multichannel service market where all providers offered the same high-priced large bundles of channels. Subscribers to these internet-distributed channel bundles still needed to pay for internet service.

A month later, Sony launched its Playstation Vue service. For $50 per month, subscribers could access fifty channels and add theme-based packages of channels similar to Sling. The service was geographically limited and available in only seven cities its first year, during which time the service

a. *Multichannel Video Programming Distributor* (MVPD) is the collective term for companies providing multichannel service: cable, satellite, and telcos. The "virtual" denoted that these services did not own the actual infrastructure used to deliver the channels.

also lacked Disney-owned channels. In its second year, Vue offered a "Slim" version of the service with fifty-five channels for $30 and content from all major conglomerates. The service could also be used on iPads. Although it didn't initially have VOD or DVR capability, favorite shows could be saved for a month and programs aired within three days could be accessed at any time. Sony delivered the service through Playstation 2 and 3 devices and touted its interface and search functionality as selling points.[2]

The internet-distributed channel bundle services didn't stay skinny for long. Although there was significant consumer frustration with the packaging and pricing of cable packages, few regarded the internet-distributed channel bundles as a good solution. Initially, loss of on-demand and DVR functionality was a significant compromise and the packages often left out some content viewers desired. Many cable companies incentivized subscribers to maintain video service by charging more for internet service without cable, which reduced possible cost savings. By September 2016, analysts estimated that Sling TV had fewer than one million subscribers, while Vue had only 100,000.[3] Nonetheless, both Google and Hulu announced plans in 2016 to bring similar services to market in the next year. By June 2017, media measurement company comScore reported an aggregate of 3.1 million subscribers across the services.[4]

The bigger impact of the internet-distributed channel bundles—as well as portals such as Netflix—was the way they pushed legacy multichannel providers to repackage their services. Verizon marketed its "Custom TV" bundle in May 2015, which wasn't particularly skinny. It claimed its base package included thirty-five channels, but it was more like thirteen common and desirable channels for $55, and $10 for additional packs.[5] In July 2015, Comcast began offering internet service and HBO for $50 and five months later tested a $15 service called Stream that included only broadcast channels and HBO in Chicago and Boston. Just as VOD content had been "impossible" for providers to negotiate before Netflix, the changing competitive environment enabled providers to negotiate different agreements with channels that allowed the providers a more competitive offering.

There was a surprising absence among all this innovation: no solution from Apple. Many expected a revolutionary announcement about Apple TV in September 2015. Short of CEO Tim Cook's declaration that "the future of TV is apps," the new version of the device launched that year wasn't nearly as transformative as the slew of portal and internet-distributed channel

Table 29.1
Major internet-distributed channel bundle services

Service	Launch	Subscribers, July 2017
DISH Sling TV	2015	1.7M
Sony PlayStation Vue	2015	450K
DirecTV Now	2016	200K
Hulu Live	2017	160K
YouTube TV	2017	

Source: Gerry Smith, "Online TV Is Growing Too Slowly to Stop the Bleeding in Cable," *BloombergTechnology*, July 25, 2017; https://www.bloomberg.com/news/articles/2017-07-25/online-tv-is-growing-too-slowly-to-stop-the-bleeding-in-cable

bundles. Apple's difficulties introducing an innovative product revealed the importance of owning content. As content holders pivoted their own businesses to encompass distribution, they were unwilling to allow Apple the same distribution stranglehold it achieved in the internet distribution of other media.

Internet distribution changed the landscape of television delivery in two ways throughout 2015. The internet-delivered offerings available from HBO, Showtime, CBS, and Noggin added to options Netflix, Hulu, Amazon Video, and YouTube had made available for viewers seeking to cobble together content from discrete services. At the same time, new services such as Sling and Sony introduced internet-delivered packages of channels that offered an alternative to a conventional multichannel subscription. These new options pressured cable providers to create more options for consumers who sought cost savings in return for cord shaving.

Such moves signaled providers' recognition of the shifting competitive space. Notably, the peril of unbundling was far greater for channel owners than for the service providers. It is worth reiterating that cable service providers delivered internet service to most homes and internet service had much higher profit margins. Transitioning customers to only internet-distributed services was not a great threat to the service providers who increased the cost of internet service for viewers ending bundling and as viewers' internet use increased as a result.

Other than Netflix—which reached 43 percent of U.S. homes at the end of 2016—none of the services offered an option that viewers stampeded toward. This was to be expected. Viewers en masse may have wanted to break out of the bloated packages cable providers offered, but what they wanted instead varied considerably. Some were driven mostly by a desire to rein in costs. The convenience of viewing independent of a schedule drove others. Some wanted that independence and an ad-free option and were willing to pay in exchange. Still others prioritized the ability to watch on an array of devices. These varying desires led to slow growth of a range of services that offered many different value propositions.[6]

The multichannel providers also reached an important inflection point in 2015: the number of homes receiving internet from the nation's largest provider, Comcast, surpassed those receiving its cable service. In many ways this was a symbolic transition. The internet was important to homes for far more than its provision of television, but this shift made visible a largely unrealized metamorphosis: internet service had grown more critical to the daily life of Americans than access to broadcast or cable television. Yet because internet service was arguably less competitive than multichannel service, the portals and internet-distributed channel bundles threatened merely to shift one version of monopoly control to another.

Past lessons suggest that fewer than half of the portals or internet-distributed bundles announced in 2015 will exist in 2020 and even fewer in 2025. Rather than introducing the companies likely to dominate internet distribution, 2015 made clear that legacy companies had been preparing to pivot to internet distribution. Future innovation would be done outside of a paradigm of television distributed through a schedule. The portals may have been the chicken or maybe the egg, but the transformation had begun. The cable industry born out of retransmitting broadcast signals into mountain valleys had become the dominant, and often monopolistic, provider of internet access.

30 Signs of Failure

Portals and internet-distributed channel bundles contributed to a sense of swift change. In between the launches of new ways to view—which, to be fair, few signed up for—by 2015, a steady stream of evidence of the faltering of legacy business models began to flow in.

Ratings for cable series across nearly all channels were notably diminished for the 2014–2015 season, but the declines also were evident across the television schedule. The ratings of programs viewed with their commercials within three days of airing among viewers aged eighteen to forty-nine—the most important metric—dropped by 9 percent in 2014.[1] The audience for live broadcast viewing was down 30 percent from the preinternet distribution season of 2008–2009.[2] In January 2015, the Nielsen ratings service reported that only 55 percent of primetime ratings came from live viewing.[3] Viewer embrace of new ways of viewing had begun to expand beyond early adopters just as the industry introduced innovative services that took advantage of the capabilities of internet distribution.

There was no clear reason that live viewing declined so precipitously. As one journalist described it, "the great media migration of American consumers to time-shifted viewing options has accelerated to a gallop this season, and it shows no sign of slowing down."[4] Alan Wurtzel, NBCUniversal's president of research explained in February 2015 that "the growth and change of (consumer) behavior is just in the early days." Just two months later he declared, "I do think we will look back on 2015 as a defining year, simply because we really have reached a tipping point."[5]

Most viewing of internet-distributed television was done through subscriber-funded services, which meant that viewing of ad-supported television was decreasing. A clever bit of reporting calculated that Netflix saved its subscribers from over 130 hours of advertisements each year.[6] The shift away from scheduled viewing was particularly high among eighteen to forty-nine year olds, but it was evident among all demographic groups.[7] Most important, it wasn't the case that viewers weren't watching. Many of the new ways of watching—including much of the growing use of VOD—wasn't counted in ratings unless the viewing occurred within a week.

Analysts blamed the decline on increased viewership of services such as Netflix. Analysis by MoffettNathanson Research argued that Netflix viewing accounted for 6 percent of total traditional television viewing in the first quarter of 2015, which amounted to 43 percent of the decline in live television.[8] Beyond Netflix, the increasing consistency and reliability of VOD offerings enticed more viewers. The measurement service Rentrak reported that on-demand viewing of broadcast series increased 19.8 percent and viewing of cable series grew 8.2 percent during 2014.[9] It wasn't that viewing was declining; what was changing was how and where viewers were watching.

Viewers were embracing new technologies for viewing that were incompatible with the established norms of revenue models built on television advertising. Subscriber-funded services such as Netflix had no advertising. Video on demand often included some advertising, but if viewed after more than a week, its advertising didn't count in the live rating.

The crisis of 2015 spilled onto Wall Street in August following statements from Disney executives that ESPN profits would be affected by subscriber losses. Conventional wisdom held that ESPN was immune to the woes experienced by the rest of television because of its reliance on sport events, which viewers still preferred to watch live. When ESPN proved vulnerable, Wall Street reacted with a swift sell-off of media stocks. ESPN laid off three hundred workers in October. Most media stocks recovered from the sell-off by the end of the year, but the perception diminished that some sectors of television would emerge unscathed.

The challenge to the traditional television business was mounted on two fronts: declining live audiences meant fewer viewers to sell to advertisers; and the small, but significant number who cobbled together content from internet-delivered sources or shifted to smaller bundles meant channels reached fewer homes, which decreased subscriber revenue.

Those trying to hold the bundle together admonished that without it some channels would be lost. But would anyone miss what was lost? Would truly desired content not simply find a more economically efficient means of distribution? Had there ever been demand for these channels, or had conglomerates simply launched them because they could? A variety of uncompetitive dynamics had long ago divorced cable from an efficient marketplace and internet distribution was providing a correction.

Many mistook the nature of the crisis introduced by internet distribution. Rather than being on the verge of death as headlines repeatedly suggested, the medium of television remained vital. The new viewing technologies and on-demand experience enabled by internet distribution, however, challenged existing practices and business operations. Practices developed for an era of television schedules were shoehorned into a changing industry as long as was possible. Eventually the gulf between norms was too great. It was no longer productive to perpetuate practices designed for a scheduled environment. The content surplus and capabilities of internet distribution undermined past strategies.

After more than a decade of forecasted death, in 2015 the contours of television's future slowly came into view. Throughout that year, much of the industry abruptly changed course, tacked into the wind, and recharted practices for an era of internet distribution. The ride ahead was bound to be bumpy. New revenue streams wouldn't replace the old ones immediately, and internet distribution on a mass scale was still untested. No one knew how quickly the audience still watching the schedule would follow course, but the tide was bound to turn.

31 A Vision of the Future

Understanding why the sea change of 2015 took so long—especially given the much earlier availability of internet-distribution technologies—requires recognizing how many separate industries and interests make up what we casually think of as a uniform television industry. The arrival of portals provided those who made television with many more places to sell series, but few had budgets for series production. Portals were not only important as a new source of television entertainment; more significantly, they also acculturated viewers to a new way of viewing and experiencing television. The many sectors of the television industry based on distribution—broadcast networks and stations, cable channels—faced increased competition. Nearly sixty years of industry norms needed to be adapted or simply forgotten in order for the business of television to accommodate the changes made possible by internet distribution. Although audiences quickly adopted new behaviors for viewing, the business of television made more measured adjustments.

The entities best positioned to lead the way were also the ones with the most to lose. The need to reinvent business strategies to account for the technological differences of internet distribution contributed to the industry's slow response, but fear of cannibalization mainly tempered evolution. Broadcast networks and cable channels maintained massive advertiser revenue despite shrinking audiences. The quiet pivots to increase reliance on subscriber funding, as well as to owning their intellectual property, buttressed them against the abrupt transformation the music and print industries had faced. But legacy practices were so ingrained that they could obscure other options and possibilities for the business of television. Eventually change could not be denied, and any attempt to slow change became more of a liability.

The legacy industry clung to legacy practices from 2010 through 2014 as a slow stream of intrepid—or just fed up—viewers found new ways to watch television and redefined their expectations of the medium. New distributors such as Netflix, Hulu, and Amazon Video entered the market early, but they didn't immediately require the television business to change to the extent that digital distribution had forced on the print and recording industries by unbundling newspapers and albums. The portals did not provide all that the television viewers desired. Consequently, many types of television are absent from the discussion here: news, talk shows, game shows, and especially sports. In time, some company would identify a way to distribute television news that took advantage of the opportunities of internet distribution, and perhaps YouTube's personalities suggested the future of television talk. Eventually, some portal would create a value proposition that similarly captured the fancy of audiences enough to shift other types of programming further from the schedule. By the end of 2016, YouTube and Facebook Live challenged those sectors of television by demanding viewers' attention with other types of programming. Internet-distributed television reconfigured a significant sector of television entertainment, but many sectors remained intact.

Although many of the industry's key decision makers and leaders hoped to escape to retirement before it became necessary to embrace internet distribution and all the transformation it introduced, by 2015 that moment could no longer be delayed. The adjustments that emerged throughout 2015 signaled a tipping point. Legacy content owners launched portals and the mounting alternatives to the cable bundle suggested unbundling had at last arrived to television.

The emergence of internet-distributed television in many ways paralleled the launch of cable, but it also was different. Cable had sought to be perceived as a legitimate part of television by replicating broadcast strategies, but internet distributors have not attempted to be like previous television distributors. Portals have built businesses that emphasize their technological difference, which makes this evolution for television more jarring.

Over the years of researching and writing this book I've debated whether this is a story of evolution or revolution. The business of television has never been static. There were periods in which the general contours of the business remained fairly stable, but adjustment—evolution—has been constant. Does the development of internet distribution seem more seismic simply because of its immediacy, or is it truly of a different order? Does it constitute a revolution?

Revolution has definitions ranging from "alteration, change" to "upheaval; reversal of fortune" to, most drastically, "overthrow of an established government or social order by those previously subject to it." To argue for a revolution in this case requires a definition more on the order of change and upheaval. Freeing television from scheduled delivery has profoundly changed the relationship of viewers to the medium and required significant adjustments in the operations of the industry. Despite that, the companies that have controlled television since its origins remain among those that dominate the industry. The programming produced today, although changed, is not radically dissimilar from the past.

The overthrow of scheduled delivery is the piece I regard as most worthy of justifying the term revolution. U.S. television is just at the beginning of developing new norms and practices in response. The prevalence of subscriber-funding among portals further shifts the business of U.S. television and the programs that it creates in ways that make it more like other media. The creation of services that offer libraries instead of schedules requires new strategies we've barely considered. The scale of adjustment that results from divorcing television from a schedule may not be fully evident, but it soon will be.

I may pin the revolution on unscheduled distribution, but it is crucial to understand that a multitude of changes was required to bring this adjustment into being. Of course, DVRs enabled viewers to use television this way a decade earlier. But adoption of this technology was slow and didn't inspire changes in industrial practices, such as releasing a whole season of episodes at once. It was not one single disruption; there is no one event we can point to and say "that's when television changed."

We often presume the new will *replace* the old, but there are also good reasons to expect that broadcast, multichannel, and internet television distribution will coexist. The different technological capabilities of broadcast and internet suit them for different uses and thus may enable a long future for both. Such a future likely requires considerable adjustments in broadcast television. This is not dissimilar to the way U.S. radio reinvented its content and reestablished a local identity after television arrived and usurped its status as provider of a nationally distributed schedule of programs; broadcast television must now identify ways to leverage its strengths. Broadcasting is a powerful technology for connecting communities based on geography—something that the business strategies of digital distribution have deemphasized. Finding a way to leverage that strength to commercial ends requires a different business strategy than creating national and global scale, which is now the norm in many media industries.

Television may never return to as uniform a set of audience and production practices as was characteristic of the broadcast paradigm. Imagining the future requires letting go of assumptions of a single victorious distribution technology. Different programming suits different distribution technologies, and different distribution technologies enable different types of programming. The dominance of a single distribution technology and technology of viewing throughout the twentieth century encouraged uniformity in programming and practice. In an environment of multiple distribution technologies with different capabilities, a wider range of industrial practices, norms of viewing, and even distinct television industries should be expected.

Conclusion: All We Need to Know About the Future of Television …

We Now Disrupt This Broadcast tells the story of U.S. television's profound evolution over the last two decades. I had been studying the changes in television for fifteen years when I started this book, but I couldn't explain what had happened. I could point to significant moments or events, some of which were clearly important at the time—as in October 2005 when iTunes first priced an episode of a television program. But many developments that seemed notable when they happened proved insignificant. Others revealed their consequences much later. It was only from the vantage point of a decade or two on that the meaningful developments became distinct from the barrage of announcements, false starts, and misdirection.

The milestones may have been well known—HBO, *The Shield, Mad Men,* Netflix—but this book explains how the milestones came to be and what the less well known reasons are to consider them as such. Some shows may have been popular or personal favorites, but they were also important because of a business adjustment occurring in the background. The behind-the-scenes changes enabled what transpired on screen and most clearly prepared the future of television.

As I pitched this project, the most consistent suggestion was to write not a book that looked back, but one that looked forward. "Explain what is going to happen in the future, like with virtual reality," a generous supporter suggested. If I could write with more certainty about that future, I would do so. Although the contours of a future of television that includes internet distribution are now more clear, opining about what is ahead would be a foolish endeavor.

The lessons of the last twenty years uncovered here, however, do suggest what is to come. There's little in the overall tale that is surprising, yet, somehow, few predicted the landscape of 2016. The lessons of the past suggest that the companies that dominate media in the current era will continue to do so in the future, except the one that unexpectedly breaks through (Netflix) and the one that fails to evolve quickly enough (Viacom?).

A new industry sector such as virtual reality (VR) allows an opening for a new company that can figure out how to provide something we don't already have and don't know that we need, *and* that can find a way to monetize it (see Facebook). But accomplishing all of these tasks is really difficult. Someone will figure it out, but many more will try and fail. There isn't a magic blueprint for the one that succeeds—in addition to finding the right concept, it is a matter of timing, execution, and luck.

The biggest challenge to successfully predicting the future of VR is that it isn't yet clear how people will want to use it. In that sense, there is a clear parallel to the internet in the late 1990s. Will it compete with other intellectual property–based media and steal our attention away from television, films, books, and videogames? Will we use it primarily to augment our interpersonal communication as the next stage of social media? Will we use it for education, or to make our surroundings more pleasant? Or will it involve some combination of these uses?

Our experience of the last twenty years should caution us to be skeptical about claims of new things "killing off" existing media and media practices. This is not to say that legacy media won't be disrupted. It isn't yet clear whether VR is a medium or a distribution system. I suspect the latter, although it could be something else entirely.

Predicting the pace of technological adoption further exacerbates the futility of prognostication. Many established media companies, such as CNN, Time Warner, and MTV, as well as myriad start-ups made exactly the right pivots toward internet-distributed video around 2005, but they were five years ahead of the viewers they were seeking. If there's a lesson to be learned there, it's one about timing. Knowing you have to time the market right is a far cry from having the tools and insight to do so. Will VR adoption be more like smartphones, DVRs, or digital television in terms of its pace of adoption? History tells us that adoption of new technology depends on many different factors, but none of them reliably predicts.

The evidence of this book also suggests that revolutions are slower and more measured than might be expected. Revolutionizing television required more than twenty years of sustained disruption. None of the futurists propounding television's future in the mid-1990s got it all right. A few had some of the pieces, but understanding a technology's capabilities doesn't reveal corresponding adjustments in business models, competitive practices, or a whole host of contextual factors that shift alongside the evolution and advance or restrain the tide of change accordingly. Just as futurists of the mid-1990s predicted, today I can order a pizza with the push of a button. But the button isn't on my remote control, as they forecast. There still isn't a way to just as easily buy the sweater an actress is wearing or a vase seen on the set, despite predictions that were almost as ubiquitous as the pizza forecast.

Although it doesn't allow us to accurately prognosticate, this book contains a lot of lessons for the future of media. In 1998, it wasn't a question of whether people would use the internet, but of when and how they would use it. So too now for VR and subsequent technology. The answer to the question of *how* remains most crucial. Mercurial beasts that we are, we had no more hope at the close of the twentieth century of predicting the time that would soon be devoted to status updates, selfie sticks, or PewDiePie than we do today of predicting how humans will use VR. Just as is the case with previous technologies, our use of new technologies will depend on:

the capabilities and limitations of technologies,
how technology companies make them usable,
our available resources of money and time,
formal and informal regulations of all sorts, and
our innermost desires and sources of pleasure.

Like other technologies, we'll use them to tell us stories, to interact with friends or strangers, and to watch animals do surprising things.

... We Can Learn from Book Publishing

When critics began comparing distinctive television series such as *The Wire*, *The Sopranos*, and *Mad Men* with storytelling in a novel, they didn't know how right they were. The comparison was meant as an indication of artistic value, but this was the first of many developing parallels between television series and novels. These series that inspired effusive praise—even among those who had long disdained television—emerged before the transformative change internet distribution encouraged. They told complex stories and deeply engaged audiences despite industry norms that largely discouraged a rich storytelling experience. Audiences could access only a "chapter" a week and often had to wait a year or more between sets of ten to thirteen chapters. *Mad Men* viewers tolerated commercial interruptions.

Parallels between internet-distributed television and the novel exist not only in distinctive storytelling, but also in competitive strategies. Those steeped in literary history could draw connections between the intermittent access broadcast norms availed and the original publication of Dickens in serialized installments in newspapers. The distribution mechanism used to transmit his stories had also changed over time. The novel's collection of chapters mirrors Netflix's drop of a full season of episodes.

Those creating these distinctive programs for cable channels were guided by a half-century of commercial broadcast practices that suggested little comparison between television and novels. Nearly twenty years after *The Sopranos'* debut, however, television could be readily experienced as the written word had been for centuries: at a viewer's own pace and on his or her schedule. The best of these series also achieved the new standard of excellence necessary well before John Landgraf and others imagined what the new standard would be. They were television shows people would watch, and even talk about, ten years after their production.

Broadcast networks and cable channels were rigorously constrained by their distribution technology. Internet distribution made television much more like other media that had not been tied to time and schedule—curiously, perhaps most like the book publishing industry.

But why would book publishing offer any insight about the future of television?

Watching a season of a television series is comparable to the time required to read a novel. The book publishing industry is accustomed to an "unscheduled" distribution environment in which every new book competes for the time and attention of readers against other new books as well as every book ever written, much as television shows increasingly do now. Publishers develop books without thought of time constraints or building a schedule. They have built their business on understanding the rhythms of consumers' leisure needs independent of underperforming time slots or coordinating the right "lead in."

Far more books are published each year than critics or readers can read; as a result, a "surplus" of books has existed for decades, if not centuries. At any given time, a commuter train will have riders reading a current release, last year's best sellers, and decades-old classic titles; television viewing in contemporary neighborhoods might be similarly dispersed. Book publishers consequently have business models based on creating and circulating content that balance revenue from new titles (new series), new content from known authors (new seasons of established series), and revenue from a backlist of titles published over the years (library rights). Of course the costs of production differentiate these industries pronouncedly. The comparison nevertheless remains helpful to break the constraining hold of schedules and time in thinking about possibilities for television.

That the future of television series could be informed by book publishing seems an unlikely irony. How could the digital wonder of the internet lead back to the conventions of the prototypical analog form?

A successful future for the business of television doesn't involve trying to force the square peg of internet-distributed television into the round hole of the norms developed for broadcast television. Competitors emerging since the advent of internet distribution need not dominate the future of television. To compete, legacy companies will have to stop trying to apply the old rules and strategies and redevelop television according to the possibilities of internet distribution. Broadcast distribution better allowed for controlling the flow of programs; audiences could be forced to wait. But do strategies of constraint lead to success in book publishing? What happens to a publisher that makes a title difficult to obtain or refuses to release books in the formats readers desire? Readers move on to another title. There are always more interesting books than anyone can read.[1]

The affordances of internet distribution have required every media industry to adjust its business model. For many media, the characteristics of analog distribution required or encouraged ownership because distribution was tied to a physical form. The only way to consistently access the written word, recorded sound, and eventually moving pictures was to buy books, magazines, newspapers, and records/tapes/CDs. Business models were based on the costs of producing and selling these goods. Owning television, while introduced by VHS technology, became a viable business only after digitization through DVDs.

Digitization and internet distribution free media from physical formats, however, and consumers have responded by indicating that—for the most part—they do not want to own media. Ownership takes up space—whether on bookshelves or as device files. Consumers understand ownership to warrant a greater financial outlay and, in most cases, the additional cost for ownership is not warranted when portals provide "rental" availability. Why own movies or television shows if portals like Netflix or HBO Now provide access, especially for media likely to be consumed only once?

Answering the questions of how to provide access and what business model provides adequate revenue for content creators and value for consumers are the next challenges of internet distribution. Although few may realize it, Netflix and HBO draw from a mostly forgotten model of book distribution—that of the circulating libraries of the 1800s. At a time before public libraries and when the cost of books made them too expensive for all but the wealthy, circulating libraries that required either a fee for membership, a per rental fee, or a combination emerged to make books—particularly fiction—available to the middle class.

Other business models exist, but media will be more vital if the efficiencies enabled by internet distribution reach creators and consumers rather than further enriching distributors. Continuing to imagine television only within existing television revenue models and business strategies excludes many possibilities when greater experimentation is needed. Other business models might be more efficient than the one Netflix originated, although many have simply followed its lead.[2]

But beyond the business of television, what is its future? Without question, the technologies available and the business models employed affect what shows are produced. This book began its story of change by exploring the behind-the-screen forces that led to the emergence of distinctive television series produced by cable channels. It wasn't that broadcasters had never thought of or aspired to produce distinctive series; rather, the advertising-based business model that developed in relation to the limitations of broadcast distribution and U.S. disregard for public service media did not encourage such programs. Cable channels had different needs. As the number of channels proliferated, a channel needed to stand out to be successful. To draw viewers from broadcast, cable programs needed to target interests and tastes more specifically than broadcasters' big tent-approach could accomplish. Cable channels' second revenue stream from subscribers allowed them to sacrifice mass advertising dollars to pursue those niche tastes.

Likewise, the rupture introduced by internet distribution provides many opportunities to redefine the content produced by the industry in extraordinary ways. Already, once unimaginable variation exists. Multiple sectors of internet-distributed television have emerged—including some largely not considered in this book. The story of the emergence of YouTube as an open access, advertiser-supported distributor would have required even more nuance in this already complicated story. By 2016, there was little evidence YouTube and the development of what were called multichannel networks (MCNs) such as Fullscreen, Maker, and Machinima were part of the story told here about long-form, scripted series and the attributes of subscription business models. The content these outlets produced—at least as of 2016—was not likely to be television that people would watch and talk about ten years hence. It may have been internet distributed as well, but it was otherwise proving to be a distinct industry. It was also so young that its logics were unclear, and it was still largely fueled by venture capital or corporate benefactors to an extent that made its future highly uncertain. This industry sector is as worthy of the close examination afforded to legacy television here, but it needs to have its own story told.[3]

Many futures of television are possible, and the businesses of television will be more multifaceted than in the past. The business of scripted series might look to the book publishing industry for clues to the future, while sports programming continues to thrive under the norms of broadcast distribution, and the YouTube/open access universe creates content of and for the moment. Netflix established a clear trajectory as a producer and distributor of television for the coming era. What remains uncertain is whether Netflix can continue to license enough content to maintain its value proposition if the studios choose to pursue their own portals. Not all will have pockets deep enough to create their own portals, so distribution businesses based on licensing series will likely persist, although it is probable that a single company will dominate.

Smart industry leaders forecast a decrease in the quantity of series produced as part of the "too much TV" discussion that emerged in 2015. Such an adjustment seems reasonable, although production will not likely decrease to an extent that a surplus of programming will cease to exist. The sense of "too much TV" emerged before legacy television outlets shifted to development strategies designed for internet distribution. In coming years, studios will reassess their practices and benchmarks of how much programming, targeting what audiences, creates the right equilibrium. Legacy subscription services such as HBO and Showtime that have been less tethered to schedule-based thinking, as well as nearly internet-native Netflix, have already begun to suggest what balance of new series, returning series, and library content will satisfy subscribers.

The content of television has already gone through extensive transformation and makes it difficult to speak of television programming in a coherent way. Further change will come after more technological innovation—perhaps virtual reality or other enhancements in image quality—but such developments are a decade from mainstream significance. Before further transformation in storytelling, portals must establish practices for building television libraries rather than schedules. These new practices will drive the storytelling of the future. Television will become something quite different as coming generations consider it not in terms of a schedule or as something watched simultaneously by a nation.[4]

Without the schedule, television will require new ways of finding, organizing, and sharing its shows. Many have focused on the roles social media and recommendation algorithms might play, but new behaviors that don't require a media tool will also emerge. Returning again to book publishing, I can't help but wonder if five or ten years from now some will begin attending monthly TV clubs as many have done with books in the past. What shows will they likely discuss? Most certainly they will be ones that are distinctive and readily accessible.

Notes

Chapter 1

1. Maureen Ryan, "*The Sopranos* Is the Most Influential Drama Ever," *Tribune News Service*, April; 4, 2007; Mary McNamera, "The Best Written TV Show: *The Sopranos*, Of Course," *Los Angeles Times*, June 3, 2013; June Thomas, "Revolution on the Small Screen: How Prestige Dramas like *The Sopranos* Transformed TV," *Newsday*, July 7, 2013.

2. In telling these two stories—of how cable channels transformed U.S. television programming and then how internet distribution revolutionized it all—this book charts the demise of the "broadcast paradigm," its replacement with a short-lived "broadcast/cable paradigm," and the development of a paradigm of "internet distribution" that now exists alongside the broadcast/cable paradigm.

I draw from Thomas Kuhn in using the term *paradigm* to indicate not only the practices and modes of behavior that surround television, but also a broad framework that includes how television is understood by television viewers and those working in its industries. [Thomas Kuhn, *The Structure of Scientific Revolutions* (Chicago: University of Chicago Press, 1970).] The broadcast paradigm was still very much in place in the mid-1990s when this book begins and includes all the dominant expectations of what television is and can be, how television is made, and what television programming looks like. Understanding television's change through the lens of paradigm shifts helps explain the slow pace of change and provides helpful conceptualization of how change happens and barriers to change in creative industries that is developed in case studies that provide finer detail.

Paradigm change is uncommon. It requires a fundamental reconsideration of what something is to occur on a broad scale rather than simply among a select few. It is often difficult to see paradigm change in the moment it is happening—and this was true for U.S. television.

The claim of a "revolution" also requires more grounding. Revolution can imply different types of change and is used just as accurately to mean "alteration, change; upheaval; reversal of fortune; a dramatic or wide-reaching change in conditions" as "overthrow of an established government or social order by those previously subject to it; forcible substitution of a new form of government," or the difference between definitions 8a and 7a listed in the *Oxford English Dictionary*. My use in this context is the former rather than the latter. This is not an ideological revolution or an uprising. It is a vast and fundamental change in a medium of communication, but not one that alters structures of power.

Reasonable and well-informed people may differ in what they perceive as the threshold of change that warrants the term *revolution*. "Television," as a medium constituted by "technologies, industrial formations, government policies, and practices of looking," as Lynn Spigel specifies, has never been static, but is always evolving. [Lynn Spigel, "Introduction," in *Television After TV: Essays of a Medium in Transition*, ed. Lynn Spigel and Jan Olsson (Durham, NC: Duke University Press, 2004), 2.] My threshold for asserting revolution derives from the scale of change. The emergence of television led to a "revolution" of radio in the 1950s, as many key industrial and audience behaviors for radio changed radically. This is now the case for television. Its revolution is not yet over. Perhaps my own sense of revolution derives from expectations not yet clearly in mass use.

The change explored in the first half of this book—of how cable channels finally came to produce original series that competed with broadcast offerings and led cable to be considered as legitimate a form of television as broadcast—is a story of transformation, not revolution. It is symptomatic of the steady change common to a medium. But the shifts that derive from the emergence of the internet as a technology of distribution are revolutionary. It is important to note that this is not yet a revolution that involves the overthrow of previous technologies of distribution that is characteristic of Raymond Williams's conceptualization of technological change being first emergent, then dominant, then residual. [Raymond Williams, *Marxism and Literature* (Oxford: Oxford University Press, 1977).] Rather, for now, and perhaps for some time, William Uricchio's conceptualization of pluriformity provides a more useful frame. [William Uricchio, "Contextualizing the Broadcast Era: Nation, Commerce, and Constraint," *Annals of the American Academy of Political and Social Science* 625 (2009): 60–73.].

The scale, evidence, and implications of the revolution are not yet clear. The continued adjustment wrought by internet distribution will continue to play out for decades. That continued change will not be other revolutions. Despite the fact that I do not advance an argument of revolution as replacement—of internet distribution replacing broadcast distribution—the emergence of internet-distributed television, the differences in its affordances, and the new protocols it correspondingly allows are characteristic of the definition of revolution noted here.

3. Cable companies provided the bulk of U.S. home internet service as of 2013; Emil Protalinski, "Over 70% of US Households Now Have Broadband Internet Access, with Cable Powering Over 50% of the Market," *The Next Web*, December 9, 2013; https://thenextweb.com/insider/2013/12/09/70-us-households-now-broadband-internet-access-cable-powering-50-market/#.tnw_pKUwZp3h

4. Susan Crawford, *Captive Audience: The Telecom Industry and Monopoly Power in the New Gilded Age* (New Haven: Yale University Press, 2013).

Chapter 2

1. Pending purchase of Time Warner by AT&T.

2. PricewaterhouseCoopers, "Big Bets for the U.S. Cable Industry: Key Opportunities for Future Revenue Growth," January 2005, 8.

3. Nick Petrillo, IBIS World Industry Report 51521, Cable Networks in the US, March 2016; Nick Petrillo, IBIS World Industry Report 51711a, Cable Providers in the US, May 2016.

4. IBISWorld Market Industry Reports, December 19, 2014.

5. Many expansive histories of cable's origins and first formations exist. [Thomas Streeter, "Blue Skies and Strange Bedfellows: The Discourse of Cable Television," in *The Revolution Wasn't Televised: Sixties Television and Social Conflict*, ed. Lynn Spigel and Michael Curtin (London: Routledge, 1997), 221–242; Thomas Streeter, "The Cable Fable Revisited: Discourse, Policy, and the Making of Cable Television," *Critical Studies in Media Communication* 4, no. 2 (1987): 174–200; Megan Mullen, *Television in the Multichannel Age: A Brief History of Cable Television* (Malden, MA: Wiley-Blackwell, 2008); Megan Mullen, *The Rise of Cable Programming in the United States: Revolution or Evolution?* (Austin: University of Texas Press, 2009).]

Cable historian Megan Mullen organizes U.S. cable history into six eras, the last of which is "Multichannel Television's Mature Years (1993–)." [Mullen, *The Rise of Cable Programming in the United States*.] Similarly, in his history, Cable Center president Larry Satkowiak organizes cable's story into three periods, with the final "Age of Innovation" starting in 1995. [Larry Satkowiak, *The Cable Industry: A Short History through Three Generations* (Denver: The Cable Center, 2015).] Our story begins in this mature stage; a time by which the cable industry had come to be a common part of the U.S. television landscape, but was still clearly a second-tier version of the television business.

Just as "television" in 2016 varied considerably around the globe, so did the history and experience of cable before the 1990s within the United States. Even in the early 2000s, substantial differences in access to multichannel television existed based largely on one's location: rural or urban, mountainous or flat, heavy tree cover and poor weather or open and clear skies. In cable history, geography determined whether a home would have cable as an option, while today it determines the likelihood and level of competition.

By many accounts, the start of the cable industry can be traced to the late 1940s, although only retrospectively can these first entrepreneurs be seen as the start of the modern industry. Relative to the billion-dollar companies that now dominate cable, the industry's origins were most modest. Cable service providers' original mission was simply to enable communities unable to receive television signals sent over the air to receive the same television channels that mesmerized their less geographically challenged counterparts. Originally called CATV (community antenna television), many of the earliest services were little more than antennas placed atop hills that carried signals to homes in the valleys below by wire. Although the industry's origins date to the 1940s, it took nearly thirty years for cable to begin to resemble the industry considered here.

A key reason for cable's slow development was the strength of the broadcast industry and its effectiveness at limiting competition. The earliest cable regulations required CATV services to carry all local broadcasters and disallowed cable service in the one hundred most populous cities. These regulations effectively prevented cable from being anything other than a retransmitter of local broadcast signals, which aided broadcasters by expanding the reach of their signals while also prohibiting cable providers from developing any service that might compete for viewers' attention. New rules in 1969 briefly forced cable systems to create local content, but this proved too expensive for the still-nascent industry, and the rules were eliminated. Other rules enacted by the Federal Communications Commission (FCC) in the late 1960s—as well as key court decisions—acknowledged the arrival of cable as an alternative mechanism for delivering video signals, but reinforced the broadcast paradigm by adopting most of its norms.

Throughout the 1960s, broadcasters were successful in perpetuating a belief among regulators that anything short of a strictly regulated cable industry would crush their businesses, but by the early 1970s, actual economic analysis emerged that refuted broadcasters' "sky is falling" narrative. The analysis happened to coincide with increasing discontent with the state of commercial television among viewers and many in government. A new regulatory framework slowly developed throughout the 1970s as cable operators eventually won the ability to import distant signals, which at last allowed an expansion in available content and enabled the business of creating channels with content independent from that of broadcasters. Indeed, other provisions ensured broadcasters' continued primary status—and certainly, only broadcast networks had the economies of scale to secure national advertisers whose fees enabled the production of the high-budget programming to which the nation had grown accustomed.

The long and contradictory history of cable regulation then veered toward deregulation in the 1980s. Congress passed a largely deregulatory act in 1984 that classified cable as "Title VI," a regulatory distinction that defined cable as a one-way video service. [Cable Communications Policy Act of 1984.] The act otherwise marked an effort to decrease cable's varied, locally based regulatory norms with uniform federal authority, although with little federal oversight. Cable service grew precipitously throughout the 1980s as a result. The approach of allowing cable to be monitored by the "market" foundered, however, because no market existed. Cable providers functioned as a monopoly and extracted yearly increases in subscriber fees that soon drew considerable complaint.

Congress subsequently reregulated cable by another act in 1992, which established myriad new rules, most notably related to rate regulation and redefining the terms requiring cable system's carriage of broadcast stations—called *retransmission consent.* [Cable Television Consumer Protection and Competition Act of 1992.] The retransmission consent ensured that cable providers either carried all the broadcast stations in their area or allowed large stations to request payment if providers wanted to include them. Such rules continued to provide regulatory support for the broadcast paradigm and will be considered with greater detail later in this book. These 1992 rules prepared the competitive equilibrium of cable service that was then disrupted in 1996.

6. NCTA, *Cable Developments, 2003* (Washington, DC: NCTA, 2013), 8.

7. Into the early 1990s, cable channels offered minimal original programming, and cable consequently posed little threat to broadcast's dominance. Merely providing an alternative to broadcast was enough to entice the majority of homes to pay for cable service. It still wasn't clear whether cable would be an ancillary business that existed alongside broadcasting or whether the two sectors would meaningfully compete. To be sure, the broadcast paradigm remained very much entrenched when this book begins its story in 1996. Cable had simply staked claims around it.

In many ways, the television landscape looked very different in 1996 than it does now, and it can be stunning to consider how much change transpired in a mere twenty years. Mass viewing was still the norm. Television was ubiquitous—found in more than 98 percent of American homes—and nearly 73 percent of homes had more than one set. Six broadcast networks reached 74 percent of those viewing during prime time in the 1995–1996 season. [Morrie Gellman, "All Bets on NBC for 1996–7," *Broadcasting & Cable,* August 5, 1996, 27.]

Subscribing to cable in 1996 typically amounted to receiving forty-seven channels in the most common "expanded basic" service, and the average monthly price for basic cable service was $24.41, or $36.81 in 2015 dollars. [Tracy Stevens, *International Television & Video Almanac: 2000*; FCC, *Report on Cable Industry Prices,* June 7, 2013.] The number of hours spent watching television was still rising in 1996 and had reached over seven hours per day per home. VCRs could be found in 82 percent of homes, and this was the early heyday of video rental. Satellite distribution was still mostly limited to the giant C-band dishes found in rural America and reached only 5 percent of television households.

Roughly half of those subscribing to cable paid to access subscription television services such as HBO and Showtime. The subscriber base of these channels was still expanding, but was larger as a percentage of television households than in 2007 when the subscription rate began again to grow after eight years of subscriber losses. Assuming no homes had both cable and satellite, 28 percent of homes received only broadcast channels, while the other nearly three-quarters were dominated by cable. Satellite largely served only those who did not have access to cable.

Although all the cable channels that are familiar names in 2016 existed in 1996, several were organized around different themes. Cable "brands" had not yet emerged—at least for general-interest channels—and most channels primarily offered an array of "acquired" programs—in other words, series that originally had been developed for and aired on broadcast networks. Only a few scripted, original cable series aired before 1996, and nearly all were on subscriber-funded cable channels such as HBO and Showtime. The most-watched cable shows of a given week included Nickelodeon's kids' shows, wrestling, and movies. The most-watched channels in both prime time and total day viewing were TNT, USA, Nickelodeon, and TBS, which all drew a little fewer than two million viewers. [John Higgins, "TNT Tops Prime Time," *Broadcasting & Cable*, December 22, 1997, 29.] By 1996 there was industry buzz about the coming of high-definition (HD), but, while the first broadcasts occurred, much like 3D in 2013 and Ultra 4K HD in 2017, only a minuscule amount of programming existed. Distribution of HD signals was most limited, and few but the earliest of adopters had ventured to buy an HD set. Significantly, the television screen was still the only screen for viewing.

8. Telling the story of cable requires establishing a few distinctions. In addition to the already noted difference between the cable channel and cable service industries, it is important to distinguish between types of cable channels. The first differentiation is between those that are advertiser supported and those that are subscriber supported. These different business models—whether the channel earns revenue from both advertisers and subscribers or just subscribers—make these sectors of the cable business remarkably different. This book tells the story of both and is careful to indicate the distinction throughout.

A second distinction is between "general-interest cable channels"—those that most closely approximated the programming norms of the broadcast networks—and "niche channels," which target narrow and specific audiences often underserved by broadcast networks or focus on a particular type of programming. Many of the most successful niche channels featured programming offered only in specific and limited time periods on broadcast networks, such as news, sports, and kids programming. Certainly, they drew away viewers who might previously have settled for whatever the broadcast networks had on offer, but the business of these channels was not a clear challenge to broadcasters.

Broadcasters focused on the emergence of new broadcast competitors such as Fox in the late 1980s and The WB and UPN in the mid-1990s—entities that competed more directly and were drawing away younger viewers. At this time, general-interest channels such as USA, TNT, and TBS featured schedules full of programs that had previously aired on broadcast networks. Even though cable channels provided minimal new content, viewers valued them for offering an alternative to broadcast networks. This was, after all, an era well before a mantra of "whatever you want to watch, whenever you want to watch it" could even be imagined.

Much of the opportunity of cable derived from the fact that television was organized by a schedule. The distribution technology of broadcasting was capable of transmitting only one signal at a time, so a set of norms—the broadcast paradigm—developed based on that technological limitation. The business of the networks involved devising and providing a schedule of viewing, referred to throughout this book as a *linear schedule*. Although cable had other capabilities, especially once rebuilt with digital architecture, it developed as a retransmitter of broadcast signals and modeled itself on the norms of the broadcast paradigm.

As cable developed during the period examined here, its growing cultural and economic value was often assessed relative to broadcast. This comparison has been—and continues to be—common in business, technology, and popular discussions of television. For many reasons, however, this comparison is not particularly meaningful: how do you compare two industries that may be based in distributing video, but are built on substantially different business models, regulatory frameworks, and programming expectations?

Part of the power of the broadcast paradigm was that any other form of television was immediately measured by its norms. The broadcast/cable horse race often compared programming: how do programs on broadcast networks compare in audience size or advertising rates with those on cable channels; how does viewing of all cable channels compare with viewing of all broadcast networks? The combined viewing of all cable channels in prime time topped that of broadcast networks in 2002, but the significantly greater number of cable channels in that composite obscured the fact that—outside of sports programming—any one cable channel rarely amassed an audience equivalent to a single broadcast network until 2012 and, even then, only a handful of shows drew more viewers than middling broadcast series. [Rick Kissell, "Cable Shows Aren't So Niche Anymore," *Variety*, November 15, 2012.]

Also, if program costs are considered, cable produced programming with much smaller budgets, which meant the channels had fewer costs to recover than broadcast networks, and there is the issue of how those costs are recovered. Ad-supported cable enjoyed subscriber revenue typically equal to the revenue earned through advertising. Cable service providers pay these subscriber fees to the cable channels each month for each household that receives the channel regardless of whether it is viewed. The dual income streams enabled cable channels' survival when faced with smaller audiences—and thus lower advertising revenue. Much of the story recounted here charts cable channels' strategic development of programming that would warrant increases in these fees paid by service providers. Subscriber-supported channels such as HBO and Showtime function under an entirely different business model that enables wholly other strategies, and broadcast networks began to receive this second revenue stream from cable service providers only in 2007.

These distinctions between sectors of cable may seem unnecessarily detailed, but these nuances are needed to understand what really happened to television and cable at the beginning of the twenty-first century. In the fuzzy memories of those who lived through it, the story is typically remembered as "*The Sopranos* changed everything, and then there was *Mad Men*, then Netflix." Indeed, these are three important milestones along the way, but they reveal little of the complicated dynamics of an amazing industrial and cultural change. We will explore the story of multiple paradigm shifts in television in detail by digging into the efforts that preceded these and other milestones and chart how evolution gave way to revolution over two decades.

Chapter 3

1. Michael Burgi, "Cable TV: A Business in Transition," *Media Week,* November. 27, 1995, 18–21.

2. Patrick R. Parsons and Robert M. Frieden, *The Cable and Satellite Television Industries* (Needham Heights, MA: Allyn & Bacon, 1998), 61.

3. A variety of cable rate regulations and rules ensuring broadcast stations' inclusion on cable systems established in legislation earlier in the decade became incorporated into standard operations by 1996. This further distinguished the cable service business from its "early days." Also, by 1996, various mergers and acquisitions significantly reorganized media ownership—these were the deals that gave broadcasters and cable channels common owners. As media scholar Jennifer Holt argues, "By the end of 1995, the transformation of media policy and structure that began in the 1980s was nearly complete." [Jennifer Holt, *Empires of Entertainment: Media Industries and the Politics of Deregulation, 1980–1996* (New Brunswick: Rutgers University Press, 2011), 163.] During the early 1990s a number of major media industry mergers, such as those between Paramount/Viacom, ABC/Disney, and Time Warner/Turner, rearranged competitive dynamics and established a norm of conglomerated media industries in contrast to the previously distinct film, broadcast, and cable television industries.

Arriving just four years after the substantial reregulation of the cable industry enacted in 1992, the Telecommunications Act of 1996 didn't change the rules of cable operation so much as it redefined the competitive terrain in which cable existed. Congress had begrudgingly enacted the 1992 regulations on the basis of consumer outcry over exponentially increasing cable rates, but it had no desire to be so extensively involved in regulation. The act attempted to increase competition by enabling companies that delivered voice services—primarily the local telephone providers who owned the wires and facilities—to enter the video service business. Many in Congress hoped that further regulatory action would not be required if a competitive marketplace could be created.

What neither the cable service nor telephone industries realized in 1996 was that landline voice and traditional cable video service would soon be technologies of the past. Within just five years, the telephone industry transitioned from growth to annual revenue losses. Indeed, linear cable video service held on longer—thanks in large part to new technologies that allowed internet protocols to be used in the transmission of video over existing cable wires. But as this book charts, within twenty years, emerging norms of internet streaming and on-demand cable viewing set the one-time core business of linear cable video service on a path of decline as well.

Of course, that future was not even imaginable as the 1996 act opened video provision to the telephone industry and allowed cable providers to offer voice service. The act cleared the regulatory hurdle preventing cross-industry competition, but was just the first step. It took a decade for the competition that was its core goal to be realized. Even then, competitive services emerged only in limited and strategic areas, and when they finally did emerge, the world of telecommunications was far different than it had been in 1996.

We will return to the video services provided by telephone companies—FiOS service by Verizon and AT&T's U-verse service—in chapter 18, when the story reaches 2004. Relevant for understanding this development as of 1996 is recognizing that the act changed the competitive norms that had prevented cross-industry competition. Even though it would take nearly a decade for that competition to materialize, the belief that competition was coming encouraged innovation in the cable industry and produced a variety of infrastructure upgrades. Cable was also encouraged to innovate because of the more immediate competition posed by satellite.

It is difficult to measure the success of the act in its aim to increase competition given the emerging technologies that shifted the playing field even more radically than this regulation might have. The fast rise of mobile telephony quickly provided a far greater threat to those companies based in landline phone service and proved an extraordinary opportunity for those who pivoted their businesses into cellular. By the early 2000s, the telephone industry had entered what economists term a "declining life stage" exhibiting "revenue growth rates ranging from –8 to –10 percent from 2002–2006." [Sylvia M. Chan-Olmsted and Miao Guo, "Strategic Bundling of Telecommunications Services: Triple-Play Strategies in the Cable TV and Telephone Industries," *Journal of Media Business Studies* 8, no. 2 (2011): 63–81.]

As a result, the opportunity to compete in home voice service proved a modest opportunity for the cable industry in terms of new revenue. Voice would be important as part of the "triple-play" bundle of video, data, and voice service that emerged once systems were upgraded to handle voice-over-internet protocol delivery (VoIP) as a standard offering by 2003. Beyond providing new monthly revenue from consumers, adding VoIP service was valuable to the cable operators because it indicated the possibility of market expansion—an attribute rewarded by Wall Street investors—at a moment of some uncertainty for the sector. The service was profitable for companies, and it also derived value because it further integrated the companies into the homes of subscribers, deterring them from changing service providers in cases where competition existed.

Clearly the bigger playing-field-shifting development that the Telecommunications Act couldn't fully anticipate in the mid-1990s was the coming importance of internet service. Both the cable and the telephone industries were well positioned to provide internet access to American homes, and at that time, there was considerable uncertainty about which provider's technology would prove superior. Telephony seemed to have the superior lines early on, but evolution in the technology used to send data over the coaxial cable that composed the cable industry's infrastructure kept this competition in flux.

4. *International Television & Video Almanac: 1998* (New York: Quigley Publishing, 1998), 20A.

5. By the end of 1996, six companies were providing some sort of satellite television service to more than six million homes. [Jim McConville and Harry A. Jessell, "Competition from the Sky," *Broadcasting & Cable*, November 25, 1996, 22–24, 23.] Eight years later, only two providers remained, DirecTV and Echo Star's Dish Network; the two proposed a merger that was denied by the Department of Justice. Despite the failed attempt at consolidation, DBS satellite introduced meaningful competition to multichannel delivery and grew its subscribers quickly.

Households in rural areas that lacked access to cable provided the primary market for early DBS subscribers, although at the end of 1995, about one quarter of satellite subscribers had canceled a cable subscription to switch to DBS. [Burgi, "Cable TV."] Although satellite had a compelling proposition at launch, it was limited until 1999 by rules originally established for large, ten-foot, C-band satellites common only in rural areas. These rules prohibited satellites from offering local stations unless the household could not receive the signals over the air. In practice, this meant that most satellite households wishing to view programming from local broadcasters would need to either erect an antenna to receive broadcast signals over the air or subscribe to the rarely acknowledged but mandated "lifeline" cable service consisting of local broadcast stations and public, educational, and government access channels that cable service providers supported in the market. Legislation in 1999 allowed satellite services to provide local broadcast channels to subscribers, which eliminated cable's main remaining competitive advantage. [Satellite Home Viewer Improvement Act, 1998.]

For those with specific programming interests, however, satellite's national footprint and digital infrastructure allowed a range of specialized services that were particularly attractive and otherwise unavailable on most cable systems. Out-of-market sports packages proved a strong point of distinction for DBS, but the service also offered packages of international and non-English channels. Even in the 1990s, cable had established "a history of service-quality problems, both with the signal itself and with the customer relationship," and DBS at last provided a service alternative. [PricewaterhouseCoopers, "Big Bets for the U.S. Cable Industry: Key Opportunities for Future Revenue Growth," January 2005, 11.] But other than this very different technology for transmission, satellite would not preserve its technological advantages for long. 6. Burgi, "Cable TV."

7. Jimmy Schaeffler, "U.S. DBS Subscriber Growth: Can Satellite's David Keep Smiting Cable's Goliath," *Satellite Today*, February 1, 2000.

8. Price Coleman, "Nowhere to Go but Up: Cable TV in 1997," *Broadcasting & Cable*, December 6, 1996, 64–78, 74.

9. Ibid.

10. Verne Gay, "Out of the Box," *Media Week*, December 1, 1997, 24–28.

11. Moreover, during broadcasters' transition to digital transmission, they were given double the spectrum space to operate both an analog and a digital signal until the transition was complete. Analog cable systems would quickly reach capacity if forced to carry broadcasters' offerings in both the old and new spectrum allocations.

12. Coleman, "Nowhere to Go but Up."

13. PricewaterhouseCoopers, "Big Bets." The deployment of digital video services—by both satellite and cable providers—was crucial to pushing the competitive conditions past that of a network era governed by the broadcast paradigm to an initial multichannel environment. Once the possibility of expanded system capacity materialized, a plethora of new channels emerged, several of which became important during the first decade of the twenty-first century, while others initially failed and entered a succession of rebrandings.

14. Data from NCTA, *Cable Developments, 2003* (Washington, DC: NCTA, 2003), 8, 11.

15. Noted by veteran Wells Fargo media industry analyst Marci Ryvicker, Internet and Television Expo, Chicago, IL, May 2015.

16. Another important development for the cable service industry during this time was the start of consolidation. The largest number of cable systems, 11,200, existed in 1993–1994. By 1996, that number had dropped to 10,000 as the cable industry increasingly sought to take advantage of economies of scale. [*International Television & Video Almanac: 2007* (New York: Quigley Publishing, 2007), 13.] An industry journalist reviewing cable ownership in 2001 noted, "merger activity was so frenzied that nine of 1999's Top 25 MSOs had disappeared into the bowels of larger operators." [John M. Higgins and Gerard Flynn, "Cable Slows, DBS Sprints," *Broadcasting & Cable*, June 4, 2001, 30–42, 32.] Of the 79 million subscribers reached by the top twenty-five operators in 2001 (including the two DBS services), the top eight providers held 88.4 percent of the market. [Figures based on data compiled by ibid., 30.] This meant an end to the days of cable service providers as local family businesses. A quick consolidation began that created the massive systems that became the internet service monopolies of the twenty-first century.

As the larger companies grew, it also became clear that there were advantages to having geographically contiguous franchises. Described by cable maverick and TCI cable system owner John Malone as the "summer of love," in the summer of 1997 operators such as TCI (which was purchased by telco AT&T in 1998, and then purchased by Comcast in 2002), Time Warner, Comcast, and Cablevision completed a series of system trades that allowed each company to cluster their subscribers and ensured that none competed with others in what media law scholar Susan Crawford terms an "unofficial non-compete agreement." [Susan Crawford, *Captive Audience: The Telecom Industry and Monopoly Power in the New Gilded Age* (New Haven: Yale University Press, 2014), 77.]

17. Michael Burgi, "Cable TV." By 1996, more than 38 percent of homes had computers and 15.2 million homes—out of a population of 97.9 million households—were using the internet, which amounted to 39.3 percent of those with computers or roughly 15 percent of American homes. The internet service that existed was, with few exceptions, the dial-up variety. Continental Cablevision launched the first cable internet service in 1994 in Cambridge, Massachusetts, although many other systems introduced service between 1997 and 1998. While technologists had a sense of the scale of possible change, at this point—and for well into the first decade of the twenty-first century—it remained a leap of faith to imagine that video could be delivered over the internet in a way that would be pleasurable for viewers.

Those who went online from home spent just thirty minutes a month doing so in the mid-1990s. There was little to look at, pages often took tens of seconds to load, and access was paid for in packages of minutes. [For more see Farhad Manjoo, "Jurassic Web," *Slate*, February 2009; http://www.slate.com/articles/technology/technology/2009/02/jurassic_web.2.html.] The first webmail site launched in 1996, and although the MP3 format had been invented, files were not yet traded. Graphical web interfaces were just being introduced and were largely isolated to offices and universities.

Although internet distribution became a central factor in television's industrial developments, it is a fairly insignificant part of the story in these early years from 1996 to 2002. High-speed data services rolled out in major markets by 1998, although only 600,000—or less than one percent of U.S. households—subscribed. The service grew quickly for cable operators, and cable had locked up about 60 percent of the total internet subscriber base by 2004.

Chapter 4

1. Kim McAvoy, "Digital Opens Door to New Nets," *Broadcasting & Cable*, January 26, 1998.

Chapter 5

1. Ed Papazian, ed., *TV Dimensions 2014* (Nutley, NJ: Media Dynamics, 2014), 75.

2. Ibid., 57. Cable ads were seen not as a bargain for advertisers, but as a liability because of the norms of the broadcast paradigm. In the 1990s, the idea of television as a medium for reaching mass audiences was still pervasive and was assumed as an inherent feature of television. It simply seemed natural that the place for big

national brands was on national broadcast networks. In time, cable channels would challenge the presumed superiority of mass audiences for advertisers' messaging by illustrating the value of more specific demographic targeting.

In some ways, parallels exist between advertisers' slow acceptance of cable advertising and the more recent measured transition of advertising budgets to digital advertising, although these transitions are more dissimilar than alike. Measuring cable audiences did not require rethinking the norms of measurement in the way digital ads on services such as Hulu or YouTube necessitated. In both cases, however, it is important to understand that the new competitor in the market threatened a massive disruption in well-established business practices. From the advertisers' vantage point, the system was not broken, so they were slow to embrace change.

3. John Higgins, "Big-Ticket Originals Pay Off for Cable," *Broadcasting & Cable* 127, no. 43 (October 20, 1997): 28–34.

4. Erin Copple Smith, "A Form in Peril?: The Evolution of the Made-for-Television Movie," in *Beyond Prime Time: Television Programming in the Network Era*, ed. Amanda D. Lotz, 138–155 (New York: Routledge, 2009), 145.

5. As cable channels became part of conglomerates that also owned studios, the conglomerates used the channels to distribute theatrical films created by their studios and created new studios to produce lower-budget films specifically for their channels.

Chapter 6

1. From 1982-1989, WTBS complemented their acquired classic sitcoms with original sitcoms that were in many ways reminiscent of those earlier series. Titles included: *The Catlins, Rocky Road, Down to Earth, New Leave It to Beaver,* and *Safe at Home.*

2. Rich Brown, "Special Children's TV Report: Cable Boosts Original Programing Slate," *Broadcasting & Cable,* July 25, 1994, 62, 64.

3. It may be difficult to see retrospectively, but the power of the dominant broadcast paradigm and the extent to which it made envisioning cable programming different from broadcast can't be overstated. Like all particularly powerful paradigms, this one drew its strength from the fact that it was invisible; broadcast norms simply were "television." How could cable come in and try to be television, but not follow the same norms? There was no U.S. precedent for deviating from all of broadcast's recognized standard conventions: seasons of twenty-two episodes, familiar genres, the annual broadcast "season," budgets of $1 to 1.5 million per hour (mid-1990s). Moreover, it wasn't clear what cable should be, or what alternative it could offer. As a result, its initial efforts looked like a cheap version of broadcast television, which unsurprisingly did little to attract audiences.

Concern about the cost of series was amplified by the uncertainty of cable business norms in the early 1990s. These first series emerged just as cable rates skyrocketed in the uncompetitive environment that sparked the need for rate regulation. Although original content was strategically valuable, cable channels soon realized that they could not continue to expect significant annual growth in the subscriber fees paid by cable providers—which were passed on to subscribers—especially as the number of channels offered expanded.

4. Other early series included, on Showtime: *A New Day in Eden* (1982–1983); *Brothers* (1984–1989); *Hard Knocks* (1987); and on HBO: *Ray Bradbury Theater* (1989–1992); *Tales from the Crypt* (1989–1996).

5. The exception is *Babylon 5*, which Warner Bros. produced for the short-lived Prime Time Entertainment Network (1993–1997) and TNT then aired in 1998. Although excluding it is certainly a contestable choice, the industry viewed this and other series produced for the Sci Fi Channel as narrowly genre focused, which is why I don't begin with it. There are also other series that predate my start. In the 1980s, a few series were shifted to cable for final seasons (*Airwolf*, USA; *Days and Nights of Molly Dodd*, Lifetime). Also, USA aired *Swamp Thing* (1990–1993), *Beyond Reality* (1991–1993), *TekWar* (1994–1996), *Duckman* (1994–1997), and *Weird Science* (1994–1998). Lifetime attempted *Veronica Clare* (1991) and *The Hidden Room* (1991–1993). The Family Channel tried *Big Brother Jake* (1990–1994), *Maniac Mansion* (1990–1993), *Zorro* (1990–1993), *The Legend of Prince Valiant* (1991–1993), and *The Mighty Jungle* (1994) and Disney developed *Avonlea/The Road to Avonlea* (1990–1996) with the CBC. Many of these early series left little record—as well as those on WTBS in the previous note—so there are likely others that I've overlooked.

6. Betsy Sharkey, "Radio Days Revival," *MediaWeek*, December 2, 1996, 16–19.

Chapter 7

1. The channel also tried to capitalize on known content by developing television versions of theatrical films, sometimes drawing on intellectual property owned by one of the studios that owned USA—Universal and Paramount. The short-lived *The Big Easy* (1996–1997) was based on the 1986 film about a homicide detective and district attorney in New Orleans, while the comedy *Weird Science* (1994–1998) was based on the 1985 Universal film.

2. The next several paragraphs are drawn from Christopher Heyn's detailed and extensive history of the series. [Christopher Heyn, *Inside Section One: Creating and Producing TV's* La Femme Nikita (Los Angeles: POV Press, 2006.]

3. Ginsberg pitched the series to Lifetime—with a first pilot written by Lydia Woodward (*China Beach, ER*)—which was developing its brand with original series. Disagreement between Ginsberg and Lifetime executives in selecting the series' head

writer prevented Lifetime from developing the series, which would have fit well with network's female-focused brand. But achieving this near deal was sufficient to motivate Warner Bros. into developing the property. A second pilot script was commissioned (for USA), and it too missed the mark. Finally, a third group of writers realized that the key to developing the story into an ongoing series was in the energy of the film's first half, of Nikita's entry into the counterterrorism organization. The first draft of this final version was completed in late 1995.

4. Rod Perth, personal communication, January 19, 2015.

5. But the production hit a stumbling block when Gaumont—one of five production entities that were cofinancing the deal—needed to withdraw because Luc Besson—director of the original *Nikita* film—decided he did not want his work adapted. Legally, the rights had been passed, but Gaumont had to take some action to maintain the relationship with Besson. Ginsberg managed to arrange for Gaumont to give *La Femme Nikita* back to Warner Bros. to keep the series alive; however, after five years of struggling to get the series made, Ginsberg, who worked for Gaumont, could not continue with the project.

6. Perth, personal communication, January 19, 2015. USA's executives—like those throughout the cable industry—realized the many benefits of original programming. Creating content that viewers had never seen before and couldn't see elsewhere brought attention to and awareness of the channel. USA could also tout its depth of original programming in subscription fee renegotiations with cable service providers held every few years in order to increase its revenue from subscriber fees. Channels may have drawn more viewers with acquired series, but being the exclusive location for original programming gave a channel particular cachet. In the late 1990s, the revenue from subscriber fees was 40 percent of the channel's revenue, and featuring original programing was likely to yield increases in advertising fees as well. Presenting quality original series would attract new viewers and more prestige, which would help close the gap between the rates advertisers were willing to pay for broadcast and those for cable.

USA wasn't alone in pursuing this strategy. As *La Femme Nikita* debuted, Lifetime prepared to reenter the realm of original programming with the drama *Any Day Now* (1998–2002) and two female-targeted comedies. With its narrower "Television for Women" brand identity, Lifetime reaped the benefits of original programming sooner than USA. The comedies it launched (*Maggie, Oh Baby*) faded quickly, but they were followed by the dramas *Strong Medicine* (2000–2006) and *The Division* (2001–2004), both of which offered a female-focused twist on conventional medical and police franchises.

Likewise, the executive team at USA had better success after *La Femme Nikita* developing original series for co-owned Sci Fi Channel with *Farscape* (1999–2003) and in taking over the final five seasons of *Stargate: SG1* (2002–2007), which had aired on Showtime since 1997. The channel also acquired the Canadian/German coproduction *Lexx* (1997–2002) and developed the original series *First Wave* (1998–2001). Targeting a more precise niche—an underserved demographic in Lifetime's case or a particular genre in that of Sci Fi—made branding easier than was the case for general-interest channels.

7. Jackie de Crinis, personal communication, April 9, 2015.

8. Heyn, *Inside Section One*.

9. Several "determinants"—many of which recur in subsequent milestones of cable's evolution and consequently become important in understanding large-scale change—are sprinkled throughout this complicated story about the real challenges of taking a creative idea and realizing it as a television series. A development process need not be as fraught as the one here, but *La Femme Nikita*'s story highlights why change is so difficult and how the broadcast paradigm was able to maintain norms for so long. For deeper analysis on this point, see Amanda D. Lotz, "Linking Industrial and Creative Change in 21st Century U.S. Television," *Media International Australia* 164, no. 1 (May 11, 2017): 10–20, special issue "TV Now," edited by Susan Turnbull, Marion McCutcheon, and Amanda D. Lotz.

10. Features of *La Femme Nikita*'s production that deviated from broadcast norms can be credited with helping the series make it to air despite its defiance of convention. Once an agreeable script and lead actress were found, USA did not require the conventional step of reviewing a pilot episode before ordering multiple episodes of the series. USA recognized that if it was going to go to the expense of producing a pilot, it might as well produce a first season. It is difficult for unconventional ideas to survive the pilot stage and the range of rethinking and second-guessing introduced. The initial order of thirteen episodes allowed the producers more narrative space to work out their ideas and find an audience.

Also, *La Femme Nikita* represented a massive allocation of USA's creative time and budget. A broadcast network could more easily look at early underperformance and afford to write it off as a failure and shelve remaining episodes because the series would be one of many in production. Once USA had spent so extensively on producing a full season, there was no real financial choice but to air it all, even when early episodes drew fewer viewers than hoped.

The fact that the series came to life at all is due to Ginsberg's persistence as a creative champion. Many unconventional series in television history owe their existence to a producer or executive who unrelentingly pushes an unlikely concept through. Creativity and creative change are enabled by industrial practices that allow for creative champions and for development norms that permit their pursuit of passion projects.

La Femme Nikita's success is also important in relation to its budget. Produced for $900,000 at a time when broadcast series budgets were typically around $1.5 million, *La Femme Nikita* was still low budget by many measures, but more expensive than many original scripted series that had been tried by cable channels to that point. The production was very frugal and the producers made sure the money it did have was on the screen rather than in executive and talent salaries. *La Femme Nikita*'s budget was low relative to broadcast, but many subsequent cable series would likewise produce series at a cost considerably below that of broadcast by deviating from norms of the broadcast paradigm—sometimes in ways that were creatively advantageous. It is important to note that *La Femme Nikita* appeared distinctive enough not to seem derivative or a cheap version of something otherwise available.

La Femme Nikita also benefited from being produced in an era in which linear television remained dominant and when the digital channel explosion had transpired for a only a small fraction of cable homes. Its debut preceded DVRs and VOD. Although series produced a decade later would benefit in other ways from such control technologies, USA had the advantage that there were still few alternatives to watching live television. *La Femme Nikita* also debuted to a much less cluttered television environment. Few other original scripted cable series were on the air when it launched, and merely being an original scripted cable series made it notable in 1997. Such a limited competitive field certainly didn't guarantee success, but it made it easier.

Likewise, the limited impact of *La Femme Nikita* can be understood by examining factors that are different for attempts at original series before and after it. Although debuting in the linear era guaranteed a captive audience, more widespread availability of control technologies would have benefitted the series in other ways. *La Femme Nikita*'s storytelling was far more episodic than serialized, but viewers who found the series later had limited opportunities to catch up on past episodes, other than the re-airing of episodes on USA. By the second season, other channels had inquired about developing deals that would have allowed *La Femme Nikita* a second airing on a broadcast network or cable channel, but USA executives decided against the risk of disconnecting *La Femme Nikita* from USA. Although selling the episodes to air elsewhere would have exposed the series to a broader audience, they believed it was too dangerous to decouple the series' identity from USA. [Perth, personal communication, January 19, 2015.]

Warner Bros.—the studio producing the series—would have profited from exposing the series in more outlets. Even this first hit reveals the contradictory interests of originating channels and studios, something that became more challenging as industrial practices shifted and encouraged vertical integration. The standard network-era practice of splitting risk and reward between the studio producing a series and the channel first distributing it became increasingly troublesome, as will be seen in several of the stories told in this book.

Merely finding a studio willing to produce an original cable series, even after a cable channel had committed to airing it, proved a consistent problem in the early days of original cable series. Studios lacked the incentive to produce content unlikely to be sold in markets after the first license. Subsequent series would eventually—and unexpectedly—prove that cable originals could be profitable in these traditional markets. And new markets such as Netflix's streaming library would eventually emerge—also unexpectedly—and prove well suited to the particularities of the series developed by cable channels. But this reality—one yet a decade off—could not even be imagined at the time.

As an original scripted series produced for a cable channel, *La Femme Nikita* struggled because it challenged the broadcast paradigm in the United States, but it also faced similar challenges abroad. International buyers knew what it meant to buy a series that had been a big hit for CBS or a mid-range performer for NBC. A series produced for an unknown cable channel required an extraordinary leap of faith. Although several features of *La Femme Nikita*'s content suggested it would be popular outside the United States, it was as difficult to convince others to buy it as it had been to produce the series.

La Femme Nikita provided an important, if barely remembered, first assault on the broadcast paradigm. Relative to later cable series, *La Femme Nikita* didn't embody the strategy of distinction that would develop more fully, but it offered glimpses of this strategy. Its creation was certainly pursued with the aspiration to change audience and industry expectations of what U.S. television could be—both as a creative form and as a business. Many of its struggles were expected, but it also revealed unanticipated ways in which the broadcast paradigm made variation or innovation difficult. Its story introduces many trials repeated in the development and production of subsequent milestones. And many of its unconventional attributes similarly proved advantageous for others.

Chapter 8

1. *OZ* provides more than just the template for subsequent subscriber-funded channel programming. It effectively pioneered the strategy that became most successful for advertiser-supported cable as well. The original series on HBO that follow *OZ* in the late 1990s were in many ways responsible for the abundance of cable series in place by 2010. Even though the strategy must be refined for an ad-supported context, the commonness of series originating outside broadcast networks and the breadth of content cable originals introduce to the television landscape trace their origin to HBO.

The term *distinction* was not used pervasively in industry discourse—at least not until after 2005. I use it to capture a strategy executives described in different ways—sometimes using "distinct" or "distinctive"—to explain the type of programming they sought in the creation of original series in the late 1990s and early 2000s. Distinctive is meant as a classification that derives from its meaning as different from, not as an evaluation suggesting "better than" (as in a fork being distinct from a spoon, though not necessarily a better tool). The term arguably follows on from what scholars earlier classified as "quality TV" in connection with the series produced by MTM productions, the independent production company headed by Grant Tinker and Mary Tyler Moore that created such critically lauded series as *The*

Mary Tyler Moore Show, *Newhart*, and *Hill Street Blues*. Achieving quality television was discussed as an industrial strategy, but it became commonly linked with a more evaluative use. [Jane Feuer and Paul Kerr, eds., *MTM: Quality TV* (London: British Film Institute, 1985).]

My argument is not that cable series were "better" than broadcast series; rather, the concept of difference from broadcast television is helpful to explain why cable originals finally succeeded as well as why this strategy became less useful by 2010. Scholarship such as Jason Mittell's examination of "complex TV" takes important steps toward identifying how these series might be argued as distinct based on textual evidence. [Jason Mittell, *Complex TV: The Poetics of Contemporary Storytelling* (New York: New York University Press, 2015)]. My use of distinctive derives from industry discourse and claimed strategy, rather than building a case based on analyzing the shows and arguing that their content was clearly different from that of broadcast networks.

2. HBO had been airing original series for twenty years before its debut, but *OZ* was unlike any previous HBO series. The channel emphasized comedy and ventured closest to drama in the coproduced anthology mystery series *The Hitchhiker* (1983–1987). HBO's original identity was primarily as a window for uncut theatrical films before the video rental market existed and for sports event programming. The distinction offered by its original series during the 1980s and early 1990s took advantage of HBO's ability to include profanity and sex, which were disallowed elsewhere on television, but otherwise brought minimal innovation.

Although the channel wouldn't use the slogan "It's Not TV, It's HBO" until 1996, it succinctly describes HBO's self-perception in the 1980s and early 1990s. Michael Fuchs, HBO CEO until 1995, conceived of HBO as deliberately outside the broadcast paradigm of television: HBO was a monthly service valued for its access to Hollywood movies. HBO did air *The Larry Sanders Show* during Fuchs's tenure, although that is explained as a result of his relationship with the series' creator and star Garry Shandling. [John Motavalli, "Albrecht: Life on the Edge," *Electronic Media*, June 17, 2002, 1, 21.] HBO's entire original programming budget was $50 million in 1990 under Fuchs, the same amount it would spend on just *From the Earth to the Moon* in 1997 after Bewkes took over. [Higgins, "Big-Ticket Originals Pay Off for Cable."]

Fuchs discouraged production of series, and the combination of original films, HBO's theatrical deals, and exclusive sports contracts established a solid business during his tenure. Indeed, in the late 1980s glory days of Mike Tyson's boxing celebrity, 20 percent of subscribers identified boxing as the reason they maintained their subscription. [Dean J. DeFino, *The HBO Effect* (New York: Bloomsbury, 2014), 62.]

3. "The Way of Success," *Broadcasting & Cable* 132, no. 45 (November 4, 2002): 6A–8A.

4. Donna Petrozzello and John M. Higgins, "HBO: Reaching for the Stars," *Broadcasting & Cable* 128, no. 12 (March 23, 1998): 28–30, 32.

5. As evident in films such as *The Josephine Baker Story* (1991), *Barbarians at the Gate* (1993), *And the Band Played On* (1994), and *Rasputin* (1996). HBO films filled the gap created once the film studios began focusing on teen audiences at the multiplex and took on topics that ad-supported networks would never broach with their made-for-TV movies.

6. Bill Mesce, Jr., *Inside the Rise of HBO: A Personal History of the Company that Transformed Television* (Jefferson, NC: McFarland, 2015), 196.

7. Ibid., 197.

8. Alan Sepinwall, *The Revolution Was Televised: The Cops, Crooks, Slingers, and Slayers Who Changed TV Drama Forever* (New York: Touchstone, 2014), 20.

9. Ibid., 20.

10. Motavalli, "Albrecht."

11. Although *Six Feet Under* might give it a good race.

12. In terms of a revenue model, *OZ* is fundamentally different from broadcast series and those created for ad-supported cable channels. HBO's subscriber funding enabled it to take the risk to defy many conventions in how shows were made under the broadcast paradigm, but the revenue model is not solely responsible; a lot should be credited to management strategy. Advertiser-supported cable channels could have challenged these norms as well.

To understand why television didn't change for so long—and why it finally did—requires accounting for a combination of business factors as well as having bold executives and organizations able and willing to take chances. The "broadcast paradigm" mostly has been used to describe a set of audience expectations that made innovation or difference difficult.

In the same way many components of the broadcast paradigm stifled audience behavior, that paradigm was also stultifying in the narrow norms it allowed in the making of shows. Features of the broadcast paradigm also governed all aspects of how shows were created and how networks licensed and managed them. HBO broke nearly all those conventions with *OZ*.

13. Of course there are regulatory differences that allow HBO to include profanity, nudity, and violence that might yield complaints or fines for a broadcast network. This regulatory difference explains the outer limits of some HBO content, but this has been used too easily to explain HBO's distinctiveness. The degree to which broadcast networks circumscribed the agency of creative talent by micromanaging them with notes is a powerful managerial strategy not inherent in the revenue model or regulatory classification.

Affording this creative freedom was intentional on HBO's part as a strategy for creating distinctive programming. As Thomopolous described, it wasn't that HBO had a policy of not giving notes, but that HBO believed in the creative vision of those who completed its development process. The common strategy within the broadcast paradigm was for the network to enforce boundaries of a series, which contributed to the consistency and homogeneity of broadcast fare. HBO built its process around letting the creator's vision take the lead, but being hands on in assisting the realization of that vision when needed.

14. Sepinwall, *The Revolution Was Televised,* 26.

15. Quoted in Steven Prigge, *Created By … Inside the Minds of TV's Top Show Creators* (Los Angeles: Sillman-James Press, 2005), 158.

16. In addition to uncommon creative freedom, it is notable that none of the myriad challenges *La Femme Nikita* faced was a source of struggle for *OZ*. *OZ*'s budget was larger than *Nikita*'s but less than a broadcast series at the time. Although its massive set was a significant investment—Fontana recalls it as the largest television set constructed up to that time—this was an expense that could be amortized over the run of the series. Setting a series in a prison also minimized the need for costly location shooting except when a story line followed one of the corrections officers home or for the backstory vignettes that revealed how each character found himself to be a prisoner at Oswald State Penitentiary.

Cost savings were achieved by actors working at rates below what they would have earned on a broadcast production [Tom Fontana, personal communication, November 3, 2015], and the series shot each episode in only seven days—uncommon for broadcast series. The complexity of the roles available aided the series's ability to attract top acting talent despite lower pay scales, and *OZ*'s shorter seasons and the fact that few characters survived all that long in *OZ* made it feasible for actors to pursue other acting opportunities as well. Some of the most significant costs of broadcast series are the fees necessary to maintain actors because series need their full availability (making it difficult for them to pursue other projects) and the extent of the year required to produce twenty-two episodes. *OZ* could allow actors greater freedom to pursue other employment.

Rysher Home Entertainment initially produced *OZ*, but the company went bankrupt during the series's run, which necessitated that Fontana find an entity to buy the program to continue production. Fontana recalls that Albrecht agreed that HBO would become the financing production partner with hardly a second thought. HBO may have happened upon the strategy of owning its original series because of this unusual situation with its first drama. This strategy enabled its bold leadership into various new forms of distribution from on demand to streaming over the next twenty years because it owned its library of original series.

Underscoring just how real HBO's risks in defying the broadcast production paradigm were, consider that a year before HBO debuted *OZ*, it announced five new comedies at the annual Television Critics' Tour: *She, Arli$$, The Corner, Another Shot*, and *The High Life*. [Alan Bash, "HBO Seeks to Fill a Wednesday Comedy Void," *USA Today*, January 22, 1996, D3.] Of these, only *Arli$$* and *The High Life* made it to air. *The High Life* was more ambitious than *Arli$$*, but it failed quickly—a valuable reminder of how uncertain risks are in the moment, even if historical distance makes the success of their innovation seem preordained.

The High Life was set in 1950s Pittsburgh, shot in black and white, and described by creator Adam Resnick as a cross between *Amos 'n' Andy* and *Glengarry Glen Ross*. [Jay Bobbin, "1950s Buddies Strive for *The High Life* in HBO Comedy Series," *Buffalo News*, November 3, 1996, TV 31.] The series aired only eight episodes and exists largely in obscurity despite the fact it has many of the same features that are part of HBO's legendary formula: Resnick originally developed the series for CBS, but brought it to HBO when he realized his creative vision wouldn't be realized at CBS; black and white shooting gave it a distinctive visual style; it was based on absurdist humor; Resnick was an established talent—he spent six years writing for *Late Night with David Letterman* and a year writing for *Saturday Night Live*, and he cocreated and wrote *Get a Life*. [Erik Adams, Sam Adams, Phil Dyess-Nugent, Will Harris, and Kyle Ryan, "It's Not TV—and It's Not Available on HBO Go: 27-plus HBO Originals Unavailable from the Streaming Service," *The A.V. Club*, May 15, 2013; http://www.avclub.com/article/its-not-tvand-its-not-available-on-hbo-go-27-plus--97744.] The series definitely "stood out," to borrow Albrecht's description of what he was seeking and saw in the *OZ* pilot. It was certainly unlike anything on a broadcast network. *The High Life* was not a drama; in that regard, *OZ* is unquestionably HBO's first. The point is that someone working at HBO in 1997 would reasonably have suspected *OZ* as likely to follow *The High Life* into the annals of forgotten series, rather than inaugurate a transformation in television.

17. Four million viewers was a larger audience than most ad-supported cable series developed before 2010. [Gary Levin, "The Inside Story on HBO's *Oz*," *USA Today*, January 2, 2003, D8.] From a business standpoint, these metrics aren't the only measure of success for a business based on gathering subscribers. Four million may have been paltry by broadcast standards at the time, but original series quickly became crucial to the programming strategy of subscriber-funded channels. Albrecht identified the value of original series as early as 1998 and explained, "We see series as a retention device, a reason to buy HBO." [Andy Meisler, "Not Even Trying to Appeal to the Masses," *New York Times*, October 4, 1998.] Likewise, Jerry Offsay, then head of programming at Showtime, was quoted in the same article explaining, "The most widely watched thing on our network is *Stargate SG-1*; it outperforms every theatrical movie on our air." [Ibid.]

These perspectives suggest a profound transformation in how original series were regarded. Just seven years earlier, in 1991, Offsay's predecessor, Steve Hewitt, opined about originals, "It's not good business. Why are we not committed to series programming in the way that we were in the past? We found that it took an inordinate amount of money." [Deborah Hastings, "Made-for-Cable Series Share Network Troubles," *Austin-American Statesman*, January 20, 1991.] Certainly the cost of original series didn't decrease, but their perceived value increased to an extent that made the costs more worthwhile.

Subscriber-funded competitor Showtime pursued a different strategy than HBO in the late 1990s. It developed original, scripted series that targeted audiences generally ignored by broadcast network fare with programs such as *Linc's* (1998–2000), *Soul Food* (2000–2004), *Resurrection Blvd.* (2000–2002), *Queer as Folk* (2000–2005), and *The L Word* (2004–2007). Although also distinctive, this programming seemed targeted to the underrepresented demographic groups represented by their casts and, perhaps as a result, did not achieve the general popular culture notoriety HBO achieved.

18. Understanding the emergence of original, scripted cable series requires appreciating the convergences and divergences of the parallel tracks pursued by advertiser- and subscriber-supported cable channels in the mid- through late 1990s. The innovation of the form occurs in conversation between them. Their different business models and regulatory structures circumscribed what each could do, but they were both outsiders to the broadcast paradigm.

It is unsurprising that the strategy of distinction emerged from a subscriber-funded cable channel—indeed, many innovations have come from such channels, as coming chapters will highlight. Certainly the corporate parent of any other cable channel could have taken the risks of deviating from the established paradigm of episode orders, allowing creative freedom, and pushing creatives to transcend boundaries. Subscriber-funded channels were more incentivized to take such risks as a matter of strategy.

Although the broadcast paradigm would be far more significantly disrupted by the developments of the early 2000s, this early period of experimentation shouldn't be overlooked. Without the failed experiments and measured achievements in the waning years of the twentieth century, the successes of the next decade might not have been achieved. The stories of *La Femme Nikita* and *OZ* also exhibit features that recur in subsequent series that further expand their accomplishments and reveal both expected and unexpected challenges in cable's assault on the broadcast paradigm.

Chapter 9

1. Cable's early original scripted series led broadcast networks to revise their focused pursuit of mass audience hits to allow some series that were not broadly popular, but appealed to narrower, yet still advertiser-desired audience segments. This couldn't be the dominant strategy of broadcasters, but it was a component of shifting strategies, especially as cable channels—in aggregate—drew away increasing percentages of viewers.

Chapter 10

1. Indeed, what makes a paradigm so powerful is the extent to which the set of norms characteristic of a given paradigm come to be inseparable from it. U.S.

television had operated completely within the broadcast paradigm for nearly four decades, and the idea that television could be different from the broadcast paradigm was consequently incomprehensible.

Internet video before Netflix is difficult to explain to those who didn't experience it, and many who did have blocked the experience from their minds. Clunky technology, fear of eroding business models, and the complexity of renegotiating licenses to include digital rights led to very little long-form, industry-produced content being available until 2010. Some content was available for purchase on Apple's iTunes, although the price point did not make this transaction—known in the industry as *electronic sell-through* (EST)—an attractive proposition for most. It is not until the latter part of the time frame considered in this section that YouTube and Hulu—still in early stages—even became part of the video ecosystem.

Harbingers of radically improved internet-distributed video appeared quickly and steadily beginning in 2007, although streaming video was a fringe behavior in these early years. The year 2010 marks the beginning of internet-distributed video in the story told here because multiple technological advancements in that year began to make streaming more than an early-adopter phenomenon. Even though it was not clear how internet-distributed video would disrupt the newly forming broadcast/cable paradigm midway into the first decade of the twenty-first century, nearly all believed some sort of disruption was about to transpire. The uncertainty about when internet distribution would become a phenomenon with mainstream implications and how the legacy television industry would respond to it also fueled the "death of television" storyline in the press.

Chapter 11

1. *America in Primetime*, "The Man of the House," The Documentary Group, 2011. Shawn Ryan interview.

2. Allison Romano, "Can *The Shield* Fix FX?" *Broadcasting & Cable* 132, no.10 (March 11, 2002): 16–17.

3. *The Shield* was not FX's first original scripted series. That credit goes to *Son of the Beach* (2000–2002), a *Baywatch* spoof produced by Howard Stern. *Son of the Beach*, along with the late-night talk show *The X Factor*, aimed to establish a younger, male target for the channel. This focus matched the audience of the major-league baseball games the channel aired, as well as the Nascar races it acquired rights to in 2001. The channel branded itself with the tagline "Fox Gone Cable" during the late 1990s as an attempt to connect the cable channel with the identity of the like-owned broadcast network, which was well known at that time as a hipper, younger skewing network home to *The Simpsons, Beverly Hills, 90210, Malcolm in the Middle, Ally McBeal,* and *The X-Files*.

In 2000, FX was described as "among the most-hated cable operators," because of the rates News Corp. demanded from cable systems to carry it. [Sallie Hofmeister,

"Brad Gray President Kevin Reilly to Head FX Entertainment Division," *Los Angeles Times,* August 17, 2000, C5.] In the early 2000s, most broadcast stations didn't require payment from cable systems to retransmit their broadcast network on the local cable system even though they were entitled to do so. Instead, the major, network-owned stations asked the systems to carry cable channels owned by the same conglomerate. This allowed conglomerates to get a foothold in cable channel competition. With digital cable becoming the norm, filling out channel capacity was preferable to cable providers than paying for broadcast networks. FX earned the distinction of being Fox's "ransom channel" in these years: in order for a cable system to carry the local Fox affiliate, the cable system had to carry FX and pay a high fee per subscriber relative to its brief existence and schedule of nearly all acquired programming. (Nonetheless, still in 2016, FX's estimated subscriber fee was surprisingly low).

The channel's president, Peter Ligouri, hired Kevin Reilly as entertainment president in 2000 and charged him with finding programs that would establish FX as the ad-supported cable version of HBO. FX had forty shows in development in 2001. Its planned first original drama was *Dope*, a look at the world of drugs from the eyes of the criminal in the vein of the film *Traffic* (2000), but it chose to launch *The Shield* instead after receiving Shawn Ryan's script. [Mike Reynolds, "FX Eyes Drama, Reality." *Cable World,*13, no. 5 (January 29, 2001): 28.]

4. Sepinwall, *The Revolution Was Televised,* 141.

5. FX had a lot riding on a series that would defy just about every aspect of the broadcast paradigm. Betting on Shawn Ryan was a significant risk. He had no experience running a show and had spent his career writing for series such as *Nash Bridges* and *Angel*. Reminiscent of Warner Bros.' hesitance to produce *La Femme Nikita,* Fox brought in Sony as a coproducer to defer some of the risk and gave Sony the international distribution rights in return. As had been the case for *La Femme Nikita,* studios were unconvinced that a scripted drama produced originally for a cable channel could be profitable even though a channel was ready to pay a license fee to have it made.

6. Jim McConville, "Arresting Bow for FX *Shield," Hollywood Reporter,* March 20, 2002.

7. As would become true of many subsequent scripted cable original series throughout the early and mid-2000s, a "successful" show was defined as much by who the audience was as by its size. Most cable series that existed for more than a season drew just under two million viewers in the eighteen- to forty-nine-year-old demographic generally most prized by advertisers. In its fourth season, *The Shield* drew about 2.25 million in this group, about 45 percent of which were women. [Anthony Crupi, "Young Blood: FX Pulls in 18–49 Demographic with Gritty, Graphic Fare," *MediaWeek,* January 16, 2016.] Most television audiences skew more female, which made *The Shield*'s audience particularly attractive to advertisers wanting to attract

the substantial male audiences watching it—men who were difficult to reach on television or expensive to reach with sports advertising.

To place this in context, in the early 2000s, the top ten broadcast shows drew ten million viewers, with the top shows drawing more than twenty million viewers of all ages (*American Idol, CSI, Friends, Survivor*). Though broadcast audiences were larger, cable series proved valuable for the specificity of their audiences. In response to controversy about *The Shield*'s edgy content, some brands pulled their advertisements from early episodes, but others joined the program in the weeks following its launch. [Jim McConville, "FX's *Shield* Hit by Ad Defections," *Hollywood Reporter*, April 10, 2002.] Brands targeting audiences likely to be attracted by the unconventional content found it a particularly valuable buy.

8. John Landgraf in Sepinwall, *The Revolution Was Televised*, 147.

9. It is important to note that *The Shield* was successful not only in branding FX and in illustrating the creative excellence that could be achieved by an ad-supported cable channel, but also in terms of commercial measures of success. To the surprise of all involved, *The Shield* was able to earn revenue for its studios in a number of additional markets. It sold internationally, adding value to the Sony library outside the United States, but it also sold more than 250,000 copies of each season on DVD. [Interview with Shawn Ryan.]

The most surprising revenue stream developed from the sale of the series to the Spike cable channel. During the same years that FX inadvertently created a particularly male-identified brand while focusing on edgy series that were distinct from broadcast fare, Spike pursued a largely unsuccessful branding campaign aimed at becoming known as the television destination for men. Spike's programmers hoped they would have better luck establishing a male-skewing brand by acquiring a male-identified series such as *The Shield*. *The Shield* also sold to local broadcast station groups, producing additional revenue for the studio.

By any measure, *The Shield* was a success—it was financially profitable, it established the FX brand, and it launched the careers of Shawn Ryan and other members of its cast (Walton Goggins) and writing staff (Kurt Sutter). Like *Monk*—the other notable original, scripted cable series debut of 2002—*The Shield*'s significant role in challenging the broadcast paradigm is often overlooked, eclipsed by the more substantial audiences or critical adulation of those series that followed. It is important to contemplate how the series landscape would be different without *The Shield*'s early-2000s intervention.

The Shield is the first of three consecutive drama hits for FX, a record still rare in original series development. FX's efforts later in the decade did not achieve as considerable success, although many were creatively noteworthy. The Steven Bochco–created *Over There* (2005) offered stories of soldiers struggling with the otherwise largely ignored Iraq War, and *Thief* (2006), starring Andre Braugher, featured an African-American protagonist in the male-centered drama. Other series were distinctive and interesting, although less successful: *Dirt* (2007–2008), a look at the world of tabloid journalism starring Courtney Cox, and *The Riches* (2007–2008), about a family of travelers who get by on grift starring Eddie Izzard and Minnie Driver. The Glenn Close legal thriller *Damages* (2007–2012) came closer to reestablishing the success of the first series, but the following year FX found *Sons of Anarchy* (2008–2014), which set record viewership both for the channel and for original, scripted cable series. *Sons of Anarchy* reasserted the channel's particular distinctive brand of series about complicated male characters that nevertheless drew substantial female audiences.

10. Significantly, ad-supported series maintained lower budgets, and the creatives that chose to produce series for them were typically paid less than they would have received for broadcast work. Cable channel's somewhat chance reinvention of the "season" as a less grueling thirteen episodes made it possible for actors to work on a series and still do other projects. It also offered actors roles with depth and complexity unlike anything on broadcast and increasingly rare in feature films.

FX also endeavored upon original comedy development in this era and had more success than most cable channels, but with as many misses as hits. *Starved* (2005), a comedy about eating disorders, quickly flopped, while the show that debuted alongside it, *It's Always Sunny in Philadelphia* (2005–) arguably became the most successful ad-supported cable comedy of the 2000s. Negotiating niche comedy tastes proved more challenging than finding niche dramas for all cable outlets. Although it wasn't difficult to find comedy voices different than those on broadcast television, finding distinctive comedy that wasn't too narrow was rare. FX's comedy about two men who primarily work as test subjects, *Testees* (2008), proved far too niche, while *The League* (2009–2015), a comedy driven by the banter among a group of friends who compete in a fantasy football league, found an audience. Precisely what distinguished the humor of all these series—successes and failures—was difficult to discern.

11. Sepinwall, *The Revolution Was Televised,* 133.

Chapter 12

1. Jackie de Crinis, personal communication, April 9, 2015.

2. Although it was the beginning of a complicated and unusual birth. ABC was still considering the series. De Crinis knew the network was also contemplating another series from Hoberman, and that it would certainly not pick up two series from the same producer (a norm that has changed considerably since then). She believed if USA was patient, it would be able to get the series from ABC with little difficulty. Redeveloping another channel's project was unheard of at this time, although it was more or less the norm by 2015. Indeed, ABC passed on *Monk* again and USA endeavored to produce a pilot. Because USA was part of a large conglomerate, it was able to contract with a co-owned studio to produce the series—saving it much of the challenge *La Femme Nikita* experienced.

3. When a series creator has little or no experience producing, it is common to have an executive production partner assigned, but Breckman refused if this producer would have final authority on scripts and threatened to walk away from the project altogether if he couldn't produce on his terms. USA eventually relented, assigning him a producer who would manage the shooting and editing of the series, which would allow Breckman to focus on writing. As had been the case with *The Shield*, the fledgling status of original, scripted cable series allowed a novice creative voice uncommon creative control.

As described by de Crinis in the first chapter, cable channels struggled to attract talent at this time because they were perceived as a backwater of second-rate programming. Cable channels trying to develop original series were unable to afford established talent. To draw talent, some channels provided an experience creatives weren't likely to have at a broadcast network. USA developed a norm of allowing creators full reign over their shows after a first season with a executive production partner. What they couldn't offer in terms of salary for creators, they made up for by allowing an extensive creative role and by providing an environment in which those who desired could quickly gain experience in all areas of series production. Though Breckman preferred just to have authority over the script, later creators such as Matt Nix (*Burn Notice*) and Steve Franks (*Psych*) took advantage of USA's boot-camp approach to gain experience in all facets of series' production.

4. The limited budget of cable series necessitated small regular casts, which led to reliance on series with single leads compared with the larger ensembles common on broadcast networks.

5. *The Dead Zone* is about a man who awakens from a six-year coma with the ability to see into the lives of anyone he touches. In a curious situation that illustrates just how difficult it is understand how media conglomerates work, conglomerate sibling channel Sci Fi also wanted *The Dead Zone* and Sci Fi and USA entered into a bidding war for the series, which USA ultimately won. Although *The Dead Zone* was not as acclaimed, it produced six successful seasons.

6. The first season of *Monk* was produced on a constrained budget, although it isn't particularly obvious. The series filmed in Toronto and had only seven days to shoot each episode, but a condition of Shaloub's contract required production move to Los Angeles if the series was renewed for a second season. Breckman noted the Los Angeles talent pool was deeper, which led to improvements in quality, and production expanded to the more common eight days per episode. In later seasons the budget was near that of a broadcast series—Shaloub proved so crucial he renegotiated his contract a few times—but there wasn't a difference in production values that most viewers would perceive.

7. USA first struggled to replicate the success of *Monk* and *The Dead Zone*, but it didn't forget the lesson that cable original series couldn't succeed if they appeared low budget. *Peacemakers* (2003), *Touching Evil* (2004), and *Kojak* (2005) were con-

sistent with other USA offerings, but they failed to draw and sustain audiences. The 2004 miniseries *The 4400* was successful enough to produce three additional seasons.

8. It is difficult to meaningfully assess how *Monk* might have performed for a broadcast network based on these outings on ABC and NBC because they were episodes that had already aired. Neither experiment was particularly successful for the broadcast networks.

9. This was similar to the success *The Shield* achieved with its sale to Spike and local stations. *Monk* achieved wider sales because it did not push content boundaries and, by being less edgy, made itself available to a broader audience.

10. Despite being very different shows, the stories behind *Monk* and *The Shield* bear a lot in common; the series established creative norms that defied the broadcast paradigm and became conventional on cable. Both series produced just thirteen episodes in their first season, unlike the twenty-two typical of a broadcast season. *La Femme Nikita* followed the broadcast norm with twenty-two (except in its abbreviated final season), but *OZ* produced just eight episodes per season (except for season four, in which it produced sixteen). *Sex and the City* varied between twelve and eighteen and *The Sopranos* consistently produced thirteen-episode seasons, perhaps offering the model both USA and FX subsequently adopted. Thirteen episodes had been a common initial order for a new broadcast series, which, if successful, would be "picked up for the back nine," extending the season to the twenty-two-episode norm. (In 2010, the ten-episode season became increasingly common and diminished season length yet further.).

Culling the season to thirteen was crucial for helping cable attract talent in its early days and can be credited with enabling the consistent accomplishment found in its series. The shorter thirteen-episode season helped offset cable's lower salaries because it allowed more opportunity for other work throughout the year. But the creative implications were likely greater than the financial ones. Countless television writers have lamented the difficulty of the demands of twenty-two-episode seasons in terms of the challenge of developing that much story and maintaining compelling narrative arcs. The shorter seasons allowed subsequent cable series to move away from conventional broadcast settings like hospitals, police stations, and courtrooms that balanced the long seasons by filling considerable storytelling with diseases/murders/cases of the week. The many different settings cable series came to explore, as well as their perceived sophistication, are in large part made possible by the shift to the shorter season.

Likewise, cable series controlled costs by developing series featuring smaller regular casts. Although this diminished the possible points of audience identification, it enabled the development of lead characters with greater depth and complexity. Cable series were able to attract top acting talent because of the uncommon roles that developed in their efforts to tell stories distinctive from broadcast.

With these volleys from ad-supported cable, which made it clear that original, scripted series competition would not remain isolated to HBO, broadcast networks responded by gradually adopting some of the practices cable initiated. The strategy of airing consecutive episodes without reruns was well suited to highly serialized shows such as *24* and *Alias* already on the air, and by 2004 and 2005, respectively, their networks shifted these shows to this practice. In time, broadcast networks also shifted to shorter seasons for series, which was intended to draw back the viewers that steadily transitioned to cable viewing.

Broadcast networks also continued to offer series that pushed the boundaries of their conventions—shows that now seemed more like the distinctive fare of cable. Such innovation had always existed around the margins of the broadcast schedule, but their designation as "cable-like" was new (and meant as a compliment). Series such as *Lost* (ABC, 2004–2010) and *Veronica Mars* (CW, 2004–2007) were as creatively innovative as early original ad-supported cable series and gave way to other broadcast experimentation that was often not as "successful" by the measures of the broadcast paradigm.

11. Mesce, Jr., *Inside the Rise of HBO,* 225.

Chapter 13

1. Amanda D. Lotz, *Redesigning Women: Television After the Network Era* (Urbana: University of Illinois Press, 2006), 42.

2. Lifetime followed the strategy of channels such as Nickelodeon that sought to be a general entertainment outlet for a particular audience. It began by focusing on health and programming similar to the morning network talk shows to create a female focus that preceded the official launch of "Television for Women" as its branding tagline in 1995. Certainly, "women" are hardly a niche. Lifetime was also advantaged by its ability to acquire series that fit its brand fairly closely because broadcast networks had always developed series with the female audience in mind. Lifetime also had a successful history of original movies. Films focused on female characters were rare in theaters and on broadcast networks, and although it also became a cliché, the Lifetime Original Movie provided a consistent yet distinctive story that likewise delivered a consistent audience.

Lifetime achieved a trio of moderately successful original scripted series in the late 1990s through early 2000s (*Any Day Now, Strong Medicine, The Division*), but it struggled with new development until *Army Wives* in 2007. Lifetime also maintained season lengths closer to broadcast norms. Its first round of series continued the twenty-two-episode norm and it backed down to eighteen episodes only for series developed well after thirteen was the norm elsewhere on cable. Relying on formulaic genres such as the cop and doc series made this episode load possible, but it also explains why Lifetime never emerged as an outlet praised for its creative excellence.

3. In prime time. Diego Vasquez, "Macho Spike TV's Quite the Lady's Thing," *Media Life*, November 3, 2004.

4. The most successful new original content during its first year was the reality series *The Joe Schmo Show*, which twisted the then still-emerging reality genre by featuring one "contestant" who thought he was part of an unscripted reality competition show while the rest of the cast was populated by actors following a loose script. *Joe Schmo* was clever and innovative, but it was not particularly consistent with "the first network for men" brand. It wouldn't have been out of place on any generally branded channel, except for constructing misogynistic versions of standard reality show contests such as "Hands on a High-Priced Hooker," in which the last contestant with a hand on a female stripper earned "immunity" from that episode's voting (a play on a popular car giveaway that required contestants to keep a hand on a car until only one contestant—who then won the car—remained). In the cases of the animated series and *The Joe Schmo Show*, Spike's original programming was not particularly related to its newfound identity as a network for men.

5. See Amanda D. Lotz, "The Impossibility of Television for Men: Lessons from Spike," in *From Networks to Netflix: A Guide to Changing Channels*, ed. Derek Johnson (Routledge, 2018) for a more extensive account and deeper reading of Spike's evolution.

Chapter 14

1. Outside the United States, alternative series are sometimes distinguished as "factual" entertainment, denoting the nonfiction aura of the series even if production techniques can extensively contrive series' basis in fact or reality. Most original cable content is best classified as *talk shows*—in which a host follows a script to explain how to do something or hosts a discussion. The category of alternative series could conceivably include talk shows, although they are not, for the most part, included here.

2. Brian Stelter, "In Cable Niches, Less Reality and More Original Shows," *New York Times*, May 8, 2011.

3. Few of the channels that developed the scripted originals discussed so far benefitted directly from the subsequent success of series or their formats in international markets or from sales to digital distributors because the studio that produced the series held these rights. But in many cases, studios owned by the same conglomerate that owned the channel produced the series, so the later revenue was valuable to the conglomerate as a whole, if not the channels directly.

4. Channels also might capitalize on the licensing of products during the quick surge in popularity some of these series experienced. Although early licensing deals

often did not include a share for the channel, reality talent increasingly had to share profits from appearances and merchandising by 2013.

5. Alternative series became common in other countries before developing widely in the United States. As the cost of acquiring U.S. dramas increased in the late 1990s, markets around the world turned to alternative series as a way to tell compelling stories at lower cost and pioneered the docusoap format.

6. Behind the scenes these shows were important to the channels because much of the risk was shifted to external production entities that produced the shows with very low margins. An unsuccessful alternative series had a minimal impact on a channel. Few resources flowed into developing it, and if it failed to find an audience, the channel simply moved on to the next thing, hoping that show might be the one to break out. Notably, the strategies behind scripted and alternative series are nearly opposed. Channels with slates of alternatives were trying lots of different low-budget ideas hoping for a flash in the pan, even if the series would likely fail within a few years. Those seeking distinction with scripted series were spending more and more, trying to stand out with excellence and originality.

Chapter 15

1. "Series," *Hollywood Reporter*, May 25, 2007; Anne Becker, "Cable Ratings Sizzle This Summer," *Broadcasting & Cable* 137, no. 30 (2007): 3; Anne Becker, "A Tough Summer Sell," *Broadcasting & Cable* 137, no. 24 (2007): 14.

2. Jon Lafayette, "AMC Hopes Ad Buyers Get 'Mad,'" *Television Week*, July 23/30, 2007, 3, 35.

3. The miniseries earned several awards and fit well with the film library that provided the majority of the channel's programming. Connecting original series development with film genres familiar to the channel suggested a feasible original strategy. AMC sought to avoid the stark brand opposition evident on a channel such as Spike, which was the *CSI* channel by day and a young men's action network by night. Although AMC's brand has seemed uncertain to many, its series are consistently "cinematic" and can be packaged with its films. *Mad Men*'s 1960s setting at least matched the era of many of the channel's films, while *The Walking Dead* most clearly provided a series extension of the horror genre.

4. Sepinwall, *The Revolution Was Televised*, 302.

5. The stories told about HBO's consideration conflict. See Sepinwall, *The Revolution Was Televised*, for a detailed discussion.

6. Lacey Rose and Michael O'Connell, "The Uncensored, Epic, Never-Told Story Behind *Mad Men*," *Hollywood Reporter*, March 11, 2015; http://www.hollywoodreporter.com/news/mad-men-uncensored-epic-never-780101.

7. Ibid.

8. A co-owned studio could be convinced of the greater good of taking on a risky enterprise with at best modest returns.

9. Most of these studios produced series that, if successful, might earn $500,000 per episode when selling in international markets and syndication, if not more. *Mad Men* seemed likely to generate just a tenth of this. By this time, DVD sales had emerged as a secondary market that could inject some revenue, but not enough to make up for the expected weak international sales. Streaming services such as Amazon Video and Netflix that would become ideal platforms for serialized narratives and pay rich licensing deals for buzz-worthy content were just developing.

10. AMC contracted with Silvercup Studios in New York to produce the pilot, and Weiner called upon the network of production talent he established working on *The Sopranos* to create a compelling illustration of what the series could be. As often the case, personal networks also played a role; Sorcher had worked with the head of Lionsgate's television unit, Kevin Beggs, earlier in his career.

11. Michael Schneider, "*Mad Men*'s True Story, from Start to Finish," *TV Insider*, March 17, 2015; www.tvinsider.com/article/906/mad-men-oral-history/.

12. "AMC Gets 'Mad' Again," *New York Post*, September 22, 2007, 61. With so much riding on the series, AMC did not risk letting it slip by in an increasingly robust environment of original cable competition. The channel spent extensively on promotion—perhaps as much as $10 to 15 million per season. General-interest channels tended to rely mostly on on-air promotion, because drawing viewers to a show on a channel they rarely, if ever, watched was costly. In addition to extensive billboard and poster purchases, Lionsgate established a deal with Macy's and AMC with Banana Republic to develop style lines and in-store displays promoting the series and connecting it with *Mad Men*–inspired clothing. Once the series and its period style became a point of cultural fascination, AMC was able to capitalize on a legion of journalists writing features about the series, but it required a considerable campaign to initially draw attention to the show.

13. Especially after a high-profile standoff between AMC, Lionsgate, and Weiner in 2011. Weiner's contract was up at the end of the fourth season, which is typically the point at which those working on a successful series try to negotiate salary increases and the studio producing the series tries to renegotiate the license fee paid by the channel to offset some of those costs. This timing made sense within the broadcast paradigm because the end of the fourth season was when a series typically had amassed enough episodes to sell in other markets. But for a cable series shooting only ten to thirteen episodes, four seasons of episodes amounted to far fewer and thus did not provide the relief of additional revenue to pay down the studio's deficit.

The sticking points of the renegotiation came down to commercial minutes and cast size. Although *Mad Men* was a very well known series, its low ratings made it difficult for AMC to monetize in conventional ways. AMC maintained significantly fewer commercial minutes per hour than most channels—a vestige of its days as a channel that aired uncut films and was subscriber supported. AMC consequently wanted two more minutes of ads in each episode.

The staff size was more of an issue for Lionsgate. Cable series managed costs by having small regular casts. Regular characters command much higher salaries than recurring characters because a regular cast role ties up actors so extensively that they are likely to do few other projects during the run of a series. But *Mad Men* had a large cast of regular characters by cable standards, and thus costs for the studio escalated quickly before it was clear how much value the series would have.

14. In prime time, among eighteen to forty-nine year olds. The Deadline Team, "2013 Basic Cable Ratings," *Deadline Hollywood*, December 31, 2013; http://deadline.com/2013/12/2013-final-cable-ratings-usa-tbs-657915

15. Jon Lafayette, "The *Mad Men* Lesson: Buzz Lights Up a Network," *Broadcasting & Cable* 140, no. 28 (July 19, 2010): 18–20.

16. Brian Steinberg, "Why *Mad Men* Has So Little to Do with Advertising," *Advertising Age*, August 2, 2010.

17. Michael Idov, "The Zombies at AMC's Doorstep: Can the *Mad Men* Network Survive Its Own Success?" *New York*, May 23, 2011.

18. Notably, the conditions and understandings of cable originals changed considerably during the eight years of *Mad Men*'s production. Just after resolving the negotiation with Weiner in 2011, Lionsgate announced a deal with Netflix in which the streaming service agreed to pay $1 million per episode for exclusive rights. [Kimberly Nordyke and Lacey Rose, "Netflix Spends $1 Million per Episode of *Mad Men* for Streaming Rights," *Hollywood Reporter*, April 5, 2011; http://www.hollywoodreporter.com/news/netflix-spends-1-million-episode-175019.] In terms of Lionsgate's economics for the series, each episode of *Mad Men* initially cost roughly $300,000 more than AMC paid. Later seasons likely had a significantly greater deficit but, by this time, Lionsgate also knew it would earn about $500,000 per episode in the international market. The length of the Netflix deal was not revealed, but even if *Mad Men* sold in no subsequent markets, the series proved extremely profitable for Lionsgate.

In reconciling this success with AMC's initial struggle to get *Mad Men* produced, it is important to recognize that no one yet imagined a deal like the one Netflix ultimately offered, and *Mad Men* seemed an unlikely property to earn so much internationally. The situation was much different for *The Walking Dead*. By the time it emerged, AMC was known internationally as the channel that produced *Mad Men* and *Breaking Bad*, which could deliver as much as 30 percent more in international fees.

Chapter 16

1. Toni Fitzgerald, "How Cable Ratings Stack Up to Broadcast," *Media Life*, March 24, 2015.

2. Cynthia Littleton, "*The Walking Dead* Powers AMC Networks' Q4 and Full-Year Earnings," *Variety*, February 26, 2015.

3. Viewers' experience of television was certainly changing as a result of internet distribution, and these changes encouraged executives at channels and networks to think about how they might lead to shifts in the business. The greater sophistication and complexity of many of the cable dramas also appealed to the tastes of international television viewers seeking programs more interesting than U.S. broadcast networks had historically produced. Moreover, it was clear by 2010 that Netflix or a service like it would be a key part of television distribution going forward. The type of series that cable channels created wasn't very valuable under previous norms of selling to broadcast affiliates that preferred sitcoms. But the on-demand viewing allowed by internet services improved the viewing experience of serialized dramas to make them much more valuable.

The internet distributors brought both challenges and opportunities. They challenged cable channels by taking away audiences. Not only did they compete for viewers during the linear airing, but some viewers began waiting for series to reach Netflix or, given the abundance of scheduled programs, simply found them there for the first time. Cable channel executives grew frustrated when hearing new viewers describe their shows as a "great new show on Netflix," because viewers did not realize the show was created for FX or AMC.

These services also brought opportunities to help series reach a larger audience, especially those that failed to break out quickly. The "Netflix effect" of bringing new viewers to the linear audience in later seasons was beneficial for some series. However, the channels that put considerable work into developing series for their schedules typically received none of the fees paid by Netflix. That revenue went to the studios. Despite the fact many original scripted series became important calling cards for the channels—establishing a channel brand and helping it stand out in the vast sea of competitors—once a show completed its initial license it became disconnected from the channel.

4. Thanks to Alisa Perren for pointing this out. Many of these differences are driven by significantly different union and guild rates for those working on them.

5. The emergence of these "cable studios" provided evidence of the changing economics of television. In an era of broadcast distribution, technology created a bottleneck that allowed only a trickle of programming through to viewers. The networks had a great business simply in airing series and selling the advertising time in them. A series would occasionally come along and earn really extraordinary profits for a studio—such as *Dallas*, *Baywatch*, *Seinfeld*, and *Friends*. But the losses on the

majority of series that never produced a second season (and thus had limited or no prospective additional buyers) encouraged the business practice of splitting costs between the studios that produced and owned shows and networks that licensed shows from studios and profited from advertising. There were also rules—the Financial Interest and Syndication, or Fin Syn, Rules, in place from the early 1970s through the mid-1990s—that prevented networks from owning most of the shows they aired.

The business model built for broadcast networks slowly eroded as competition from cable drew away millions of viewers from broadcasters' schedules throughout the 1990s. They tried to preserve their businesses by conglomerating with cable channels and other media entities throughout the 1980s and 1990s. By the time the rules preventing networks from owning their shows were eliminated, the big five media companies (Disney, Viacom, News Corp, NBC (GE), Time Warner) owned studios, broadcast networks, and cable channels, as well as many other forms of media. The multiple revenue streams of these conglomerates helped offset the consequences of the changing competitive environment.

Practices such as licensing only the first airing of series to mitigate the risk of failure—a key concern under the broadcast paradigm—became a lower priority. The competitive environment now allowed an abundance of programming. DVRs and on-demand technology gave viewers new ways to manage their viewing. The emergence of internet distribution—which was just starting to take shape in 2010—changed the macroeconomics of media again. New outlets such as Netflix sought to license series and made owning series more valuable. Money-minting broadcast blockbusters grew more rare, and new ways of viewing encouraged the development of different types of shows. Following a television mystery series that unfolded over months—even years—of scheduled television was difficult, but services such as Netflix offered a way for viewers to focus on a series from start to finish. Also, by relying on subscriber funding, Netflix measured success differently than those focused on amassing audiences for advertisers.

Analysts noted that the cost for channels to get into the studio business was relatively small. Mostly, it added roughly 30 percent to the cost of a series—the cost over the license fee that the external studios had previously borne. [Jon Lafayette, "Cable Nets Step Up Their Studio Game," *Broadcasting & Cable* 145, no. 17 (May 4, 2015): 6–8.] Many channels already had an infrastructure of development executives adequate for the few series most sought to produce. Allowing channel and studio development execs to wear two hats also streamlined the production process, creating one set of notes for creatives and ensuring the studio and channel understood the series in the same way. Mostly, though, creating their own studios was about business strategies and the increasing uncertainty of advertising revenue at a time when viewers were willingly paying subscriptions to access content on their own terms.

6. Lacey Rose, "The New Cable Model: Why It's Better to Own Than to Rent," *Hollywood Reporter*, October 12, 2010. The studio touted revenue-sharing models that

rewarded their talent, but lower guild rates for cable productions also aided in cost reduction. As is the case with cable dramas, cable comedies tend to have smaller casts, fewer episodes, and irregular seasons. These attributes diminish costs, but they also make it necessary for talent paid per episode to pursue additional projects.

7. Lafayette, "Cable Nets Step Up Their Studio Game," 7.

8. Keach Hagey, "AMC Bets on Payoff from Its Original Shows," *Wall Street Journal*, May 29, 2014.

9. The creation of cable studios was not without consequences for other parts of the industry. In many ways, the cable studios replicated the norm that developed at the broadcast networks after the elimination of rules preventing them from owning series two decades earlier. At that time, many in the creative community decried the change in practices that emerged as several independent studios—those not owned by a company owning a network—were quickly shut out of the business as networks bought most of their shows from co-owned studios. Many of these independent studios had been responsible for some of the greatest hits—both commercially and artistically. Creatives also complained about increasing bureaucracy when networks bought from co-owned studios; they found that once they all had a common employer, fewer executives functioned as creative champions.

The reproduction of the vertical integration of studios and channels in cable will likely produce similar consequences. Although there are more buyers of television content than ever before, the emergence of cable studios makes difficult the business of studios that are not connected to a network or channel. The new cable studios claim that they will not produce exclusively for their channels, but if the last two decades of the broadcast business are any indication, willingness to buy from studios that are not co-owned will be rare.

Chapter 17

1. Louise McElvogue, "U.S. Cable Firms Find Tough Crowd Abroad," *Los Angeles Times*, August 27, 1998.

2. Janet Stilton, "US TV Networks Seek to Grow by Expanding Internationally," *Adweek*, October 12, 2014.

3. Brooks Barnes, "Prime Time Ambitions," *New York Times*, August 28, 2011.

4. Paige Albiniak, "Making a World of Difference at CBS." *Broadcasting & Cable* 143, no. 3 (2013): 10–11.

5. The transition of the domestic and then global television business from one based on licensing content to distributors to one of creating and self-distributing content had consequences for creative labor. The norms of creatives' contracts were based on the assumption that series would be sold to the highest bidder when offered in

subsequent markets. A lawsuit filed by *The Walking Dead* creator Frank Darabont against AMC illustrated such concerns. [Bill Carter, "*The Walking Dead* Creator Is Suing AMC over Payouts," *New York Times*, December 19, 2013.] Top talent often receive a percentage of revenue, so that if AMC chose to "pay" itself a lower license fee when airing the series on one of its international channels than another network would have paid, the earnings of talent with "profit participation" were reduced. The studios faced several lawsuits for such practices in the 1990s when they sold series produced for broadcast networks to co-owned cable channels. Such concerns underscore how standard labor practices were affected by evolving distribution strategies. These concerns were most acute for those working under contracts crafted within the legacy norms of U.S. television.

6. Although Netflix was primarily recognized as a movie streaming business, its increasing development of original series also revealed its potential as an international television distributor. Netflix introduced changes to standard license agreements paying more for series up front but often offering talent no back-end potential because Netflix's series were unlikely to be sold through multiple windows. Its use of a cost plus fee structure eliminated the concern identified in Darabont's lawsuit against AMC, as talent were paid higher initial rates than if licensing the series under legacy practices, but they also had no possibility of later profits.

Chapter 18

1. Derived from ITVA 2012 data. Digital cable continued its rollout, increasing from 18.5 million homes in 2002 to 45.5 million by 2010. Moving cable subscribers to more expensive digital tiers allowed cable service providers to recoup some of the costs invested in the technology and expanded their system capacity. The increased image quality and expanse of channels was competitively necessary as satellite grew its subscribers faster than cable through 2010.

Cable service providers also introduced video on demand as a typical feature of digital cable service. About 7.7 million homes could access VOD in 2002, and 40 percent of television homes had access by 2010. Despite its availability, content remained most limited. As will be discussed more in chapter 23, subscriber-funded services such as HBO and Showtime were the only channels to make extensive programming available on demand. The extra access helped subscribers feel there was more value in their subscription, which was crucial given the monthly fee. Ad-supported channels were more cautious with their offerings and often lacked the rights to make programs available because studios owned the content. Audience measurement practices were not yet designed to count the audience of advertiser-supported content on VOD, so without a way to derive value, VOD remained limited to mostly pay-per-view and subscriber-funded content through 2013.

2. In terms of other benchmarks of a medium in transition, television improved qualitatively and expanded quantitatively. This is an era of considerable adoption of

high-definition television. Less than 1 percent of homes watched HD quality video in 2002, but 54.2 percent did so by 2010. Competition among channels more than doubled from 224 national cable channels in 2002 to 565 in 2010, although not all homes received such a range. The average number of channels received by a household increased from 100 channels in 2003 to 151 by 2010. [ITVA 2003, 2012, 2013.] Despite the growth in channels available, the number of channels viewed per home each month remained at roughly fifteen.

3. Digital cable clearly introduced more competitors just among traditional cable channels, but competition from "new media" video providers remained minimal in this period. Computers were the first screens to offer internet-distributed video, although nominal long-form content was available—rarely from authorized sources—and images were of poor quality and loaded slowly. Netflix began a streaming service for PCs as early as January 2007, although it offered only about 1,000 titles and subscribers were entitled to stream only one hour per month per dollar they paid for their subscription plan. The service reached a broader range of early adopters in October 2008, when streaming expanded to reach Macs, TiVos, and Samsung Blu-ray players, but this remained very much a leading edge-behavior. Even viewing YouTube videos remained a niche practice at the tail end of this era.

Apple released the iPhone in 2007 and the iPad in 2010, which laid the foundation for high-resolution screens on mobile devices, which would soon become very important, but their impact on video viewing—especially long-form, industry-produced content—was initially very minimal. By 2010, viewers watched only three hours of video per month online—or six minutes a day—and viewed 3.5 hours monthly on mobile devices. The great majority—97 percent—of video was still viewed on a television screen. The minimal use resulted mainly from the limited long-form, professional content available. By 2010, as new screen technologies such as tablets became available, those who could afford them adopted these internet-connected devices. Technology alone wasn't enough, however; services were needed to organize and distribute to these screens, and these services didn't develop substantially until after 2010.

The ubiquity and portability characteristic of television in the era of internet distribution didn't become significant until after 2010, when Netflix launched its app enabling streaming on phones and tablets and smart TVs and related devices began to make the living room screen a site for accessing internet-distributed content. Though the arrival of sophisticated internet distribution in 2010 marked the dawn of a new phase of television competition, all of these changes were not significant enough to trouble the business model of U.S. television. Business norms remained unchanged through 2010 even though the array of content viewers chose among and how they viewed it began to change tremendously.

4. Evidence of coming change was certainly noticeable before 2010, the point at which this story shifts from cable's transformation to adjustments wrought

by internet distribution. Technologies that disrupted the linear norms of the broadcast/cable paradigm, such as DVRs, delivery systems such as VOD, and the emerging array of portable screens, were too nascent for the industry to have developed strategies before 2010. Early adopters' use of such technologies—particularly DVRs—introduced anxiety in the industry, but too few used these technologies in ways that disrupted industry norms for them to yet affect industry practices. The story of disruption brought by digital technology emerges tentatively here in preparation for its more significant—yet still far from mainstream—role in the next section.

The DVR was an extraordinary development in terms of a first strike against linear viewing. It was also particularly important to cable channels and their efforts to create original series because DVRs helped manage the growing abundance of viewing options that made it increasingly difficult to keep track of when shows were "on." The capability for "season pass"–type DVR recording was also beneficial as the notion of a consistent September through May "television season" eroded and cable seasons became especially haphazard.

To early DVR users, slow adoption of the devices was puzzling. The penetration of DVRs expanded in the United States from a mere 1.5 percent in 2002 to 40 percent in 2010. They provided an easy tool for nonlinear viewing and avoiding commercials. The main increase in the momentum of mainstream adoption resulted when cable service providers began offering cable boxes with integrated DVR technology. This made acquisition and setup simpler for viewers—one fewer box to connect. DVRs were also important for cable service providers as a new service that allowed an additional fee that became a crucial revenue stream. A congressional study in 2015 found that the average U.S. household paid $230 per year in set-top box rental fees, which yielded twenty billion dollars a year for cable and satellite providers. [Brian Fung, "You're Paying a Shocking Amount to Rent Cable Boxes Every Year," *Washington Post*, July 30, 2015; https://www.washingtonpost.com/news/the-switch/wp/2015/07/30/the-average-american-pays-a-shocking-amount-to-rent-cable-boxes-each-year/; *New York Times* Editorial Board, "Let Consumers Use Better Cheaper Cable Boxes," *New York Times*, August 31, 2015; http://mobile.nytimes.com/2015/08/31/opinion/let-consumers-use-better-cheaper-cable-boxes.html.]

DVRs were important for introducing nonlinear viewing, which was a piece of the broadcast paradigm that remained relatively undisrupted by the transition to a broadcast/cable paradigm. Indeed, the fact that the business model of broadcasters and advertiser-supported cable depended on linear viewing contributed to an ostrich-like stance in an industry trying to deny coming nonlinear possibilities. Business-model adjustments in 2007 acknowledged the increasing presence of DVRs when advertising agreements began to be sold on the "C3" measurement, which counted live viewing as well as DVR viewing within three days so long as commercials were not fast-forwarded.

DVRs were disruptive to industry practices and produced a range of "sky is falling" narratives because of their implications for ad-supported television. The industry may have recognized that the expanded array of channels and abundant original offerings made the linear schedule an increasingly insufficient organization for television, but it desired a "solution" far less disruptive to business models and practices. The industry preferred the practice of repurposing, which was briefly prevalent in the early 2000s.

It is difficult to meaningfully assess the implications of the DVR alone. Certainly, the device became important to viewing practices. For some, viewing became completely divorced from the linear schedule upon the DVR's arrival, but that was not the case for most. Had the DVR emerged without internet distribution close on its tail, the old model of linear viewing may have remained the industry focus much longer.

It is also likely that the emergence of an abundance of original cable series—as transpired in the early 2000s—would not have been so successful without the nascent availability of control technologies like the DVR to assist in managing viewing. By the time linear viewing was clearly in ruins in 2015, it was impossible to extricate the role of DVR viewing from the broader fragmentation of viewing across streaming and on-demand services that emerged after 2010.

Other technologies that quickly became central to watching television were also available in these years.

5. Nielsen Media Research, "State of the Media: Usage Trends: Q3 and Q4 of 2010," 2011; Nielsen Media Research, "State of the Media: DVR Use in the U.S.," December 2010; http://www.nielsen.com/content/dam/corporate/us/en/newswire/uploads/2010/12/DVR-State-of-the-Media-Report.pdf.

6. ITVA, 2003.

7. 17.6 million households to 34.2 million, ITVA, 2012.

8. Despite this growth, the two major satellite services, DirecTV and Dish, attempted to merge in 2002, but were prohibited by the FCC from completing the deal. Competition within the satellite industry was particularly valuable for rural areas that remained without access to cable. Allowing a satellite monopoly would have eliminated the competitive marketplace for those rural subscribers, and both services survived without the merger. Satellite services continued to siphon off subscribers throughout the decade despite the cable providers' advantage of being able to bundle video and internet service. This was not as valuable until the need for high-speed and high-capacity internet became a priority—which happened once services such as Netflix made available a significant library of professional video content. The pressure from satellite foundered after 2010 when cable's ability to bundle internet and offer more advanced on-demand services became a meaningful distinction between the services.

Though cable's monopoly power was diminished by satellite, competition from telcos proved far less significant than many expected. The two largest telcos, AT&T and Verizon, found it expensive to upgrade their wires. And, as the new satellite competitors had experienced in the late 1990s, the telcos were forced to pay high carriage fees to populate their services with the expanse of channels necessary to be competitive. This made it difficult for telcos to undercut incumbent cable service providers on subscriber rates because the telcos often paid more for the same programming. The discrepancy in fees became public when AT&T bought DirecTV in 2015. DirecTV paid about $17 less per subscriber per month than AT&T because of higher rates telcos faced when they built services with little negotiating leverage. [Mike Farrell, "AT&T's Air War," *Multichannel News*, September 28, 2015: 10–16, 12.]

The opportunity for cable providers to compete in telephone services had less value than was the case before the ubiquity of mobile phones. Nevertheless, in 2008, VoIP contributed 30 percent of the profit margins earned by Charter, Comcast, and Time Warner—the largest cable-based multichannel providers. [SNL Kagan, SNL Kagan Multichannel Summit, presentation by Robin Flynn, November 11, 2013, 10.]

9. Peter Svensson, "Verizon Winds Down Expensive FiOS Expansion," *USA Today*, March 26, 2010; http://usatoday30.usatoday.com/money/industries/telecom/2010-03-26-verizon-fios_N.htm

10. By 2015, telco service had grown to 39 percent in this market, which made it one of the most highly telco-subscribed markets in the country. [SNL Kagan Special Report, U.S. Multichannel Subscriber Update and Geographic Analysis. Data from Q4 2010.]

11. With less than six million subscribers between them, by 2010 it seemed unlikely that telcos would introduce a significant challenge to cable's dominant hold on the multichannel marketplace. However, the AT&T acquisition of DirecTV and proposed acquisition of Time Warner (still pending at press) suggested a possible competitive reconfiguration that could alter the dynamics established during the twenty years explored here.

The Telecommunications Act was not a complete loss; the mere threat of coming competition encouraged—if not forced—the cable industry to innovate. In many ways, those innovations are what prevented the telcos from mounting a more significant threat and positioned cable to occupy such a dominant position in internet service. By 2015, cable lost another 6 percent to telcos, while satellite remained steady at 33 percent of the multichannel service market.

12. Cable companies maintained dominance and provided internet service to 54.5 percent of subscribers in 2010. [Based on calculations from ITVA cable internet figures: 43M cable broadband subscribers in June 2010; 78.5M broadband households.]

13. John B. Horrigan and Maeve Duggan, "Home Broadband 2015," *Pew Research Center*, December 21, 2015; http://www.pewinternet.org/2015/12/21/home-broadband-2015/.

14. One of the most profound adjustments in the economics of cable service during the first decade of the twenty-first century had nothing to do with cable channels. After seeing cable channels develop original programming, demand higher carriage fees, and impinge on their paradigm, broadcasters became increasingly envious of their dual revenue streams from advertising and carriage fees. Broadcasters even had the legal footing to access that dual stream.

One component of the cable regulation from 1992 that sought to establish the competitive field among broadcasters and cable channels included provisions regarding "must-carry" and "retransmission consent." In short, broadcast stations could be guaranteed access on the area cable system if they invoked their right to "must-carry." This provision was designed to make sure small, often independent, stations weren't prevented from reaching cable homes as more and more households found it convenient to dismantle antennas and receive broadcast stations via their cable provider. The right to retransmission consent fees was available to stations that did not request must-carry; cable providers desired to include major stations to provide a better service for subscribers. The logic was that cable services would be made stronger by having the then still dominant broadcast networks among their offerings. This legislation established that broadcast stations could expect remuneration for making their signal available to cable systems.

Curiously, for well over a decade that remuneration involved no cash payment. By the early 2000s, as digital cable and its expansive offerings reached more homes, many of the largest stations that were network owned requested carriage of new cable channels rather than a carriage fee. For example, in 2002, ABC used retransmission consent negotiations—deals between a local station and the cable provider that were typically renegotiated every few years—to launch its Toon Disney and SoapNet channels rather than request a fee from providers for its ABC-owned ABC stations. Likewise, News Corp. secured greater carriage and fees for FX by not requiring fees for News Corp.–owned Fox stations.

The largest, most valuable stations weren't requesting payment, which made it difficult for smaller stations to do so. When they did, cable providers consistently refused, afraid of the precedent it would establish. As broadcasters transitioned to high-definition and digital signals that made it possible to use some spectrum for a second channel—often used for weather or news—some were able to negotiate carriage of these digital channels. Other times they negotiated for better channel position.

Notably, satellite services did pay broadcasters, and so did telcos. This was a result of their different market power as latecomers. Satellite was initially prohibited from carrying broadcast stations, and not having the stations was a significant competitive disadvantage in trying to woo established cable customers. When the rules were changed, satellite paid for the stations and passed the costs along to subscribers. Likewise, when telcos came into the market to try to steal cable and satellite customers, they had to offer as expansive an array of programming as the other services to be competitive. Broadcast stations were able to demand significant retransmission fees because the telcos were desperate to have broadcasters on their service.

The established norms for retransmission consent changed tremendously beginning in 2006. Mostly, finances started to tighten for broadcast stations and the value of creating additional cable channels lessened. Cable channels and new broadcast stations had eroded the extreme dominance of legacy broadcast networks. Advertisers were still paying top dollar for broadcast time because a single broadcaster substantially exceeded the audience reach of any single cable channel, but the multiplicity of cable channels drew away more and more viewers each year. The mid-decade financial crisis, which decreased advertiser spending across all television providers, also pushed broadcasters to finally fight for carriage fees.

CBS—recently decoupled from the Viacom cable channels and the most-watched broadcast network—first demanded retransmission fees. Because different stations had contracts due for renegotiation in different years, norms didn't change all at once; but once stations began holding out for retransmission fees, norms shifted quickly. Cable providers paid $160.7 million in retransmission fees in 2008. [Tim Doyle, "Retransmission Consent to the Rescue," *SNL Kagan Media and Communications Report*, June 3, 2009.] That total increased to $1.24 billion dollars just two years later, and fees continued to escalate as dramatically in the years covered by the next section of the book. [National Association of Broadcasters, www.nab.org/documents/newsRoom/pdfs/072213_Kagan_broadcast_cable_fees.pdf.]

In 2006, just 18 percent of subscribers were subject to retransmission consent fees, but by the time a full three-year cycle passed, cable providers were paying fees for 92 percent of subscribers. [SNL Kagan, *Broadcast Retransmission Fee Projections 2006–2016*, in Steven C. Salop, Tasneem Chipty, Martino DeStefano, Serge X. Moresi, and John R. Woodbury, "Economic Analysis of Broadcasters' Brinkmanship and Bargaining Advantages in Retransmission Consent Negotiations," June 3, 2010; https://ecfsapi.fcc.gov/file/7020499521.pdf

In different ways, the retransmission fees both slowed and hastened the cable revolution. The "found money"—as a *Variety* article described it—bought the broadcasters a few more years of the status quo. [Cynthia Littleton, "Free TV's Found Money," *Variety*, February 22, 2010, 1, 37.] The broadcast business consequently weathered the economic crash of 2008 and was able to continue with very little innovation into the mid-2010s. The combination of this new cost of broadcast channels and the steady hikes in the payments broadcasters sought pressured the economics of the cable bundle. The hue and cry for internet-delivered services resulted from the perceived lack of value of giant bundles of channels that were consumers' only option. After sports programming, broadcast retransmission fees were responsible for a considerable amount of growing programming costs.

The changes in technologies, service providers, and economics that transpired in the first decade of the twenty-first century were significant, if not as obviously revolutionary as those that came soon after. The cable infrastructure upgrade was invisible to most in this era, but of enormous importance. The first decade of the twenty-first century was a time of considerable uncertainty that derived from a limited sense of what the internet would be and how it would be important to television. Most viewed it as a threat, but this uncertainty about the internet, combined with anxiety caused by the shift from a broadcast to a broadcast/cable paradigm, created a perception of television in peril. Any night of the week, though, that sense of peril was challenged by the abundance of television content available, much of which expanded viewers' perceptions of what television could be—in a good way. It was a decade of contradictions.

The retransmission fees paid by cable providers to broadcast stations were not intended to produce broader change. Retrospectively, though, this can be identified as the first substantial crack in the foundation of traditional bundled channel delivery. These fees, which were costly for the services and passed on to subscribers, overburdened the already stressed practice of selling access to channels in bloated service tiers.

15. ITVA, 2007. Of those subscribing to high-speed, internet service, roughly two-thirds subscribed to a cable company for service, while one-third purchased DSL from a telco. [Mike Farrell, "Cable's New Double Play," *Multichannel News* 31, no. 4 (October 25, 2010): 10–11, 23.] What is important is that only 33 percent of all consumers had a choice between these services in 2002. [Enrico C. Soriano, Lee Tiedrich, Amy Levine, and Emily Hancock, "A Look at Key Issues Currently Shaping Internet Deployment and Regulation," *Computer & Internet Lawyer* 21, no. 7 (July 2004): 1–22, 2.] Also important, at this time 200 kb/s was the distinction required to qualify as high speed.

16. Anonymous, "The Internet Was Kind of Terrible in 2002," *WebPro News*, August 16, 2012; http://www.webpronews.com/the-internet-was-kind-of-terrible-in-2002-infographic-2012-08

17. In 2008, the FCC increased the standard of high-speed to 768 kb/s, and then to 4 mb/s in 2010.

Chapter 19

1. By myriad measures, the broadcast/cable paradigm replaced the broadcast paradigm by 2010. The fading inertia of broadcast and surging cable development suggested that the trajectory of U.S. television might even have become a cable/broadcast paradigm.

2. The story recounted here is not the first case of the fundamental remaking of a cultural industry, nor will it be the last. By digging past the easy and

most-remembered story, this book explains why change was possible at certain points and just how the redefining changes came to be. Understanding the nuances of this transition is what allows television's experience to inform understandings of how media industries change. See Lotz, "Linking Industrial and Creative Change in 21st Century U.S. Television" for an article that takes such analysis as its focus.

Chapter 20

1. Netflix, "Netflix Launches U.S. Subscriptions Plan for Streaming Movies & TV Shows Over the Internet for $7.99 a Month," *Netflix Media Center*, November 22, 2010; https://media.netflix.com/en/press-releases/netflix-launches-us-subscription-plan-for-streaming-movies-and-tv-shows-over-the-internet-for-dollar799-a-month-migration-2

2. iOS in 2010; Android in 2011.

3. Other services too, but earlier internet required much buffering.

Chapter 21

1. Of course people did pay for television in terms of cable and HBO subscriptions, but I assert this was conceptually different from DVD purchase. Netflix provided the most substantive push into internet-distributed television and led to the rethinking of the medium of television in almost every way. Technologically, the affordances of internet-distributed services freed television from the one-to-many norm of broadcast and the bottleneck of content created by the linear schedule. Netflix provided an alternative experience of television viewing—although it changed the experience in so many ways it is difficult to know what part of the experience was most compelling. Netflix blended content historically made for ad-supported channels with a subscriber-supported revenue model. This latter attribute proved particularly important in an environment of streaming services that sought to replicate television's ad-supported norms.

Even in its DVD-by-mail days, Netflix offered a valuable way to provide access to series because it didn't mandate a rental period or charge late fees, but this was obviously just the beginning of Netflix's foray into television. Although it may have been focused on changing how audiences watched films, Netflix just as significantly upended television.

2. *The Hollywood Reporter* Staff, "TV Executives Roundtable," *Hollywood Reporter*, August 10, 2016; http://www.hollywoodreporter.com/video/watch-thr-s-full-tv-918281

3. Trefis, "Exploring Netflix's International Ambitions," *Stockpickr*, March 15, 2011; http://wwwexploring.stockpickr.com/netflix%E2%80%99s-international-ambitions.html.

4. Netflix posed the biggest challenge to cable channels' extensive reliance on acquired programming throughout much of their program day. Why would a viewer want to watch an old episode of *Law & Order* on a cable channel that filled the episode with sixteen to twenty minutes of commercials if Netflix offered the same episode without commercials and provided the convenience of starting and stopping the episode as desired? As much as 70 percent of the programming on the top ten cable networks was composed of acquired programs as late as 2015, and Netflix threatened the ample advertising income channels generated during those shows and the fees studios earned selling shows to cable channels. [Paige Albiniak, "Upfront Deals Reveal New Rules of Distribution," *Broadcasting & Cable* 145, no. 19 (2015): 27.] Early Netflix viewers were not heavy daytime viewers, so audiences for cable's acquired content remained intact through 2015. But evidence of greater changes coming was impossible to avoid. This erosion of cable channels' value as a buyer for library content was of concern to studios that had long relied on domestic syndication windows for profits—$23.5 billion a year (compared with $2.8 billion from international sales) as of 2006. [John M. Higgins, "American TV Rebounds Worldwide," *Broadcasting & Cable* 137, no. 37 (September 18, 2006): 18–19.] The *Wall Street Journal* reported that revenue for a drama after its original license was composed of 41 percent international licensing, 34 percent U.S. broadcast licensing, 17 percent syndication/reruns; and 8 percent portal licensing. [Amol Sharma, "Dramatic License: TV Studios Tune in Foreign Markets," *Wall Street Journal*, November 20, 2014, A1.]

5. The service wrote several licensing deals when it had fewer than six million subscribers, but as its subscriber base quickly grew, the studios looked more closely at the service and reassessed the value of what they licensed. Most notably, in its 2011 renegotiation for content from Starz, the subscriber service demanded a fee that recognized Netflix's subscriber growth and the danger that distributing through Netflix posed to its business model. Disagreement over fees led to the end of carriage of Starz's programming.

Such changes were an inevitable part of transitioning from a start-up outsider to a central force in television distribution. But the next stumble was self-inflicted. Netflix inelegantly handled substantial changes in its service when it announced that it would split the streaming and rental by-mail services in September 2011, a decision from which it ultimately backpedaled. Netflix was punished in the short term with subscriber losses—this also coincided with license renegotiations like the one with Starz that subscribers experienced as lost content—and some Wall Street reevaluation, but Netflix was preparing to pivot the service once again. Having decimated the legacy companies of the video rental business in which it began, Netflix began distancing itself from its identity as a separate category of service provider, announced plans to produce original programming, and began describing itself as most like subscriber-supported channel HBO.

6. Although Netflix was much less profitable than HBO because of its greater spending on programming. Netflix ramped up its content production quickly. By 2014, internet distributors including Netflix offered twenty-seven original scripted series, up from four in 2010, and the number grew to forty-four in 2015. These series competed for viewers' attention alongside a still-expanding broadcast and cable marketplace. The internet services were simply taking a page from cable's playbook and trying to connect distinctive content with their brand, but these new competitors heightened the challenge of achieving distinction even more.

As evidence of coming disaggregation of television became inescapable by 2015, it was difficult to characterize precisely what role Netflix played. Without question, it hastened the arrival of disaggregation as the first service to effectively distribute long-form programming to an array of devices. Even before most viewers owned these devices or knew they wanted to watch in a different way, Netflix created the demand for change among viewers that forced the legacy entities to evolve. Without Netflix's streaming offerings and viewers' embrace of them, cable service providers would have delayed embarking on initiatives such as TV Everywhere, and cable providers would not have developed robust video-on-demand offerings. The industry needed an outside competitor to spur it to change. [Mesce, *Inside the Rise of HBO*, 228.]

7. Mike Snider, "Those Netflix Binges Are Bad News for DVRs," *USA Today*, April 18, 2017; https://www.usatoday.com/story/tech/talkingtech/2017/04/18/nielsen-more-homes-stream-than-use-dvrs-and-netflix-top-service/100617374/

8. Rani Molla, "Most Americans Can Get Internet on Their TV—But They're Still Mostly Watching Plain Old TV," *Recode*, May 10, 2017; https://www.recode.net/2017/5/10/15611654/americans-watching-tv-streaming-online-video

9. Todd Spangler, "Netflix Caused 50 Percent of U.S. TV Viewing Drop in 2015," *Variety*, March 3, 2016; http://variety.com/2016/digital/news/netflix-tv-ratings-decline-2015-1201721672/

10. Behind NBCU, Disney, 21st Century Fox, CBS, and Time Warner, but ahead of Viacom, Discovery, A&E, Scrips, AMC, Univision, and Crown Media. [Matthew Ball and Tal Shachar, "After TV: Video's Future Will Be Bigger, More Diverse and Precarious Than Its Past," *ReDEF*, March 9, 2016; https://redef.com/original/after-tv-videos-future-will-be-bigger-more-diverse-precarious-than-its-past]

11. Statista, "Leading Reasons Why Netflix Subscribers in the U.S. Subscribed to Netflix as of January 2015"; data reported in eMarketer from a Cowan & Company study, methodology not specified; http://www.statista.com/statistics/459906/reasons-subscribe-net ix-usa/.

12. Wayne Friedman, "Most TV/Video and Mobile TV Usage Is in Home," *Media Post*, June 3, 2013; http://www.mediapost.com/publications/article/201657/most-tvvideo-and-mobile-tv-usage-is-in-home.html.

13. Jon Lafayette, "With New Deal, Viacom Steadies VOD Ship." *Broadcasting & Cable* 143, no. 23 (2013): 12.

Chapter 22

1. The reality and significance of viewers going "over the top" was overhyped and presented too simplistically at first. Although increasing amounts of video could be accessed online, it was hardly a perfect replacement for the programming most audiences chose to watch on television sets. The dire predictions also failed to acknowledge that legacy television content providers could pivot their businesses to take advantage of internet distribution. It made little sense to do so before absolutely necessary, and executives were tight-lipped about plans. Their silence fed a perception that plans were not being made, in turn making the doomsday predictors all the more vehement.

In these early years of internet-distributed television, OTT was predicted to cause waning cable subscriptions and disaggregation, but the reality during these years was far less dire than the forecasts. The forecasts weren't wrong, but they didn't prove to be right on the mark either. Like any transition of this scale, the embrace of internet-distributed video developed gradually—particularly that which came at the expense of cable-delivered video. Consider that the just-nearly-completed shift to high-definition television took well over a decade, and that change required only that viewers buy a different set. Shifting to internet-distributed content required more technical intrepidness than most viewers possessed—especially before 2013. Moreover, for those satisfied with legacy television offerings—especially major sports contests—there wasn't a significant upside to switching to internet-distributed services. Enterprising media technology reporters performed cord-cutting experiments and, through 2014, their verdicts consistently acknowledged that while they could find plenty to watch relying on internet alone they couldn't watch all of what they wanted, which led most to recommend maintaining the status quo.

Just as the period from 2002 through 2010 is important as a distinct era marking cable's infiltration into the broadcast paradigm, the conditions from 2010 through 2014 prepared the substantial disruption that arrived in 2015, when an array of internet-distributed portals and packages provided a meaningful alternative to broadcast and cable distribution. This backstory provides necessary context to the later adjustments that might otherwise seem startlingly sudden.

2. Diane Mermigas, "IP Promises Video-to-go as Next Big Media Wave," *Hollywood Reporter*, October 11/17, 2005, 8, 76.

3. Leslie Ellis, "Portable Video: Why Cable Isn't Toast," *Multichannel News* 27, no. 3 (2006): 28.

4. Legacy industry ventures that included Warner Bros.' in2TV, CBS's Innertube, CNN's Pipeline, and Comedy Central's MotherLoad tested the waters with early

internet-delivered video experiments. Most lasted only a few years. Netflix was not yet streaming, and Hulu had not yet begun operating; YouTube was in its first year, and iTunes would strike its first major television deal with ABC in October.

Chapter 23

1. Comcast's Xfinity service suggests what this might have looked like. After Comcast purchased NBCUniversal—and gained ownership of its program libraries—it included much of the back catalog in its on-demand service.

2. Two barriers prevented VOD from offering a value proposition significant enough to compel viewers to use it. First, the system needed the type of shows viewers typically watched on television, such as prime-time dramas, comedies, and unscripted series. Without this programming, subscribers seeking a nonlinear television experience instead relied on DVRs, even though VOD had the greater technological potential for watching programs on customized schedules. The "need" for VOD grew increasingly acute as the programming environment transitioned into abundance and then surplus.

The second—and related—barrier was the need for ad-supported networks and channels to derive value from making content available on demand. Early VOD advertising technology was quite limited: the advertisements had to be embedded in the program so that the same ad would air as long as the content was available and was the same for every viewer, which severely limited its usefulness. Most ad-supported channels and networks depended on advertising for at least half of their revenue. Viewers who watched on VOD past its first three days were not counted in the audience that was sold to advertisers and thus did not generate advertising income. Technology that would allow "dynamic ad insertion," the ability to change ads over time or to target particular viewers with particular ads, was in development throughout the early 2000s. Without it, VOD lacked a revenue stream, which discouraged its use by advertiser-funded channels.

3. This expanded the "multiplex" strategy subscriber-supported channels had used since 1991, which offered multiple linear channels (such as the east and west coast feed) as well as themed channels (such as HBO2, HBO Comedy, and HBO Family) when cable providers had adequate capacity.

4. The VOD sector divides content into three categories: free, subscription (e.g., HBO), and what was at first called MOD (movies on demand) —but later became TOD (transaction on demand). This latter category predominantly consists of movies, but also includes some sporting events (boxing, wrestling, MMA) and any other programming that requires a specific fee to view.

5. The fees demanded by network-owned broadcast stations for retransmission grew rapidly, starting at $.25 per household per month, but quickly escalating into deals

requiring increases to multiple dollars per household over the course of a few years. Cable providers pushed back and negotiated for rights for additional use of content they did not have before, such as the ability to stream content to subscribers' tablets, smartphones, and computers and to gain access to the most recent episodes of prime-time series for VOD delivery. As deals among cable providers and content groups were completed, the patchwork of offerings grew more consistent and plentiful.

6. On-demand service was included in the most common "expanded digital" subscription package.

7. George Winslow, "Finally, the Year of the VOD Ad," *Broadcasting & Cable* 143, no. 42 (2013): 26. Part of the reason for cable providers' success in expanding the standard digital rights of their agreements with content creators was that dynamic ad insertion technologies became capable of reaching more households by 2013. They were also able to disable the fast-forward capability to ensure that viewers allowed ads to air. But the measurement company Rentrak's VOD use data showed that only 28 percent of VOD transactions of free television entertainment programming took place within three days of airing. [Rentrak, "State of VOD, FOD: Q2 2015," Chart 18.10, 43.] By 2015, channels were increasingly able to negotiate with advertisers to include viewing within a week of airing, although this time frame still excluded more than half the on-demand transactions. In 2015, channels increasingly convinced advertisers to count viewing within seven days, what was termed the *C7 measurement.*

After those three or seven days, dynamic ad insertion allowed channels to replace and reduce the advertising and sell that advertising based on the number of views as long as the content remained available because Rentrak was able to provide census measurement of VOD use. Dynamic ad insertion allowed the ads to change and was also "addressable," meaning different ads could be sent to different households, much like ads on internet-distributed services. This capability was extremely important to cable-distributed video as it began competing with internet-distributed, ad-supported video, although a wide range of experiments continued through 2014 with few standard practices emerging. Changing the ads available allowed a new revenue stream separate from the advertising sold in the live airing.

Cable providers may have had to negotiate hard to secure digital rights for VOD, but many channels found they were forced into a strategy that proved unexpectedly valuable. Most executives at both the studios and networks assumed they would lose some of the linear audience if they made shows available outside of the linear-scheduled airing. But DVRs had already revealed some surprising behaviors. Viewers with DVRs watched more television because of the convenience offered, and VOD had the potential to provide even more convenience. Channels that made content available on demand also quickly realized that the expanded availability could help viewers who initially missed the first episodes of a show catch up with it and join the linear audience. By 2013, journalists started writing about the "Netflix effect" to describe the unusual growth some series achieved season-to-season as word of mouth spread and audiences were able to view a show from the beginning before joining the linear audience. [Andrew Wallenstein, "Netflix Flexes New Muscle with *Breaking Bad* Ratings Boom," *Variety*, August 12, 2013; http://variety.com/2013/digital/news/netflix-flexes-new-muscle-with-breaking-bad-ratings-boom-1200577029/.] Of course, not all content profited equally, with the advantage again going to content that was distinctive.

Many in the legacy television industry increasingly viewed VOD as a survival tool. Because fast forwarding could be disabled, VOD better ensured ad viewing and thus combatted DVR use, which enabled commercial skipping, while dynamic ad insertion helped television match the value proposition digital ads offered advertisers.

The industry faced more than the emergence of VOD capability, however. The technological affordances of internet distribution were more wide ranging and internet-distributed services such as Netflix also relied on a different business model and business practices. Many of the emerging internet-delivered services were subscriber supported and thus ad-free and allowed viewers access to full seasons of programming. The new internet-distributed services also didn't have to counteract the effect of decades of poor customer service and many had user interfaces far superior to what many cable providers offered. These distinctions made it difficult to understand changing audience behaviors and to know whether it was the ad-free viewing experience, self-determined viewing, or the content that was fueling adoption of internet-delivered services despite expanding VOD availability.

What was incontrovertible was that the landscape of television offerings was expanding—possibly developing multiple sectors that weren't governed by the logics that supported the industry in previous decades. VOD matched the easier organization of viewing that internet-distributed services provided as viewers became acculturated to watching anytime and in new ways—like enjoying an entire season of episodes over a week or two. By the time VOD offerings became worthwhile in 2013, 49 percent of homes had DVRs, which allowed viewers to create their own "on demand" schedule and to fast forward through commercials. [Amanda D. Lotz, *The Television Will Be Revolutionized*, 2nd rev. ed. (New York: New York University Press, 2014), 60.] In 2014, FX president John Landgraf reported that the network lost about 40 percent of potential ad revenue as a result of DVR use and commercial skipping, which propelled some channels to expand VOD offerings (in which the ability to fast forward could be disabled) and develop their own nonlinear services in which they could better control and derive data regarding viewers' behavior. [Jon Lafayette, "Original Programming Costs Raise Concerns," *Broadcasting & Cable* 144, no. 3 (2014): 8–10, 10.]

8. Just as Netflix's encroachment encouraged the development of useful VOD offerings, Netflix continued to shape VOD by reducing the licensing fees it would pay for

shows that had wide VOD availability. By 2015, cable providers increasingly secured the full current season of episodes for their VOD libraries, after which the episodes typically became unavailable because the studios would sell them to a second domestic buyer—increasingly internet-distributed services such as Netflix, Amazon, and Hulu. The fee these buyers would pay decreased if the series had extensive exposure through weeks and months on VOD platforms.

The negotiations for VOD content revealed the incompatibility of the curious and competing dynamics of historical industry norms with the new era of internet distribution. Video on demand disrupted the balance of risk and reward between channels and studios that was effective in a network era limited by the scarcity of a linear schedule. In licensing a show, a channel did not pay the full cost of production, just about 70 percent of the cost. The studio covered the deficit and retained ownership so it could sell the series in other markets to recoup that deficit. The ad-supported business models encouraged networks and channels to make it as easy as possible for viewers to see a show, or to find it late and join the live audience. But studios had different interests and were pressured by buyers for those later windows—such as Netflix—that weren't willing to pay as much for series whose episodes had all been available the previous year on a VOD platform. Studios were thus motivated to decrease VOD availability. To make this dynamic more confusing, in many cases the studio and channel were owned by the same parent company.

Co-ownership between studios and channels had often been enough to achieve the deals to allow five-episode stacks, but once studios saw revenue potential diminished, co-ownership became less persuasive. The vast media conglomerates evaluated the productivity of their studio and channel divisions separately, incentivizing the studios to prioritize making deals that ensured their divisional profitability over ones that would aid a like-owned channel. Valuing VOD—especially as dynamic ad insertion and norms for splitting advertising revenue with cable providers remained nascent—was difficult and VOD revenue was less certain than the licensing dollars internet-distributed services offered. But because internet-distributed services didn't share viewing data, studios also didn't know if they were undervaluing their content in these deals.

Moreover, the value of VOD was not the same for all types of programs. Significant differences in VOD use and pricing emerged for different genres (comedy versus drama), storytelling strategies (serialized versus episodic), and original outlet (cable channel versus broadcast network). Serialized cable drama benefited most from VOD availability. Half-hour comedies were relatively more lucrative in syndication sales, so studios sought to preserve exclusivity and allowed minimal VOD availability. As viewers became accustomed to nonlinear viewing and VOD offerings grew more extensive and consistent, the advertising in VOD streams quickly became lucrative. As early as the spring of 2013, ABC's head of sales, Gerri Wang, acknowledged that 3 percent of the prime-time audience that ABC sold to advertisers was viewing on demand. [Brian Stelter and Amy Chozick, "Viewers Start to Embrace Television on Demand," *New York Times*, May 21, 2013.]

9. In many ways it was unclear whether viewers were eschewing all ad-support or whether the problem was that the channels tried to shoehorn previous norms of linear advertising onto the VOD platform. Practices of selling, buying, and counting linear audiences developed over sixty-plus years for a distribution technology that enforced scarcity, but retrofitting these norms onto nonlinear viewing platforms was awkward and ignored the fact that audiences now could select less ad-heavy alternatives. As dynamic ad-insertion technology advanced, VOD commercial breaks grew longer and channels more frequently required cable providers to disable fast forwarding. From a viewer perspective, the prevalence of advertising and disabling fast-forward technologies ensured the continued relevance of DVRs and diminished the value proposition of VOD. Channels could have preserved the value of VOD by creating a space of less advertising clutter and selling ads at a premium. Without clarity of new practices, channels awaited a move from advertisers while advertisers sought norms from channels.

10. Rentrak, "State of VOD, FOD: Q2 2015," 3.2, p. 6, 9.2, p. 22, 10.2, p. 24. Within free-on-demand viewing, television entertainment dominated with nearly 50 percent of the transactions, compared with 24 percent for kids' programming, 14 percent for music, and the remaining 13 percent shared among movies, news and information, life and home, sports, events, and specials. Notably for the story about cable's rise as a supplier of original programming, cable-originated series were steadily viewed at least twice as frequently as those of broadcasters, and in some months the margin was over 5:1.

11. A rich account of the origins of TV Everywhere is Cliff Edwards, "Revenge of the Cable Guys," *BloombergBusinessweek*, March 11, 2010; https://www.bloomberg.com/news/articles/2010-03-11/revenge-of-the-cable-guys

12. Notably, the availability issues weren't the cable service providers' fault, but few viewers understand the intricacies between service providers and channels.

Chapter 24

1. Shalini Ramachandran, "More Cable Companies Take TV off Menu," *Wall Street Journal*, October 3, 2014.

2. Ibid.

3. Shalini Ramachandran, and Keach Hagey, "Dropping Viacom Costs Suddenlink Few Subscribers," *Wall Street Journal*, February 24, 2015.

4. Cynthia Littleton, "Cable Under Fire," *Variety*, November 4, 2014.

5. Shalini Ramachandran and Joe Flint, "ESPN Tightens Its Belt as Pressure on It Mounts," July 10, 2015.

6. Todd Spangler, "Who Are the Winners and Losers in Pay TV's Unbundled Future?" *Variety,* November 4, 2014. Ratings service Nielsen changed its measurement protocol in 2013 and expanded its definition of a "television household" to include those that accessed video only through an internet-distributed service. Nielsen reported that such homes, what it termed *zero-TV homes*, had grown from two million in 2007 to five million in 2013. Although Nielsen's measurement provided the reassuring clarity of knowing over-the-top-only viewers accounted for just over 4 percent of the population, almost half the zero-TV homes were under the age of 35, reinforcing concerns about cord nevers. [Nielsen, "Free to Move Between Screens: The Cross-Platform Report, Q4 2012," March 2013.]

7. 2014: 100.4 million; 2015: 96.6 million; 2016: 94.7 million; Todd Spangler, "Pay TV's Pain Point Gets Worse: Cord-Cutting Sped Up in 2016," *Variety*, April 14, 2017; http://variety.com/2017/digital/news/cord-cutting-2016-pay-tv-research-ott-1202030814/

8. Jonathan Hurd, "TV Viewers Love Beards—and Giving Cable a Shave," *Multichannel News* 35, no. 3 (2014): 27.

9. Keach Hagey and Shalini Ramachandran, "Pay TV's New Worry: 'Shaving' the Cord," *Wall Street Journal*, October 9, 2014; http://www.wsj.com/articles/pay-tvs-new-worry-shaving-the-cord-1412899121.

Chapter 25

1. Madeline Berg, "Why the *Game of Thrones* Audience Keeps Getting Bigger," *Forbes*, July 14, 2017; https://www.forbes.com/sites/maddieberg/2017/07/14/why-the-game-of-thrones-audience-keeps-getting-bigger/#dbff0ba57035

2. Josef Adalian, "*Game of Thrones* Is Now HBO's Most Watched Show Ever," *Vulture*, June 5, 2014; http://www.vulture.com/2014/06/game-of-thrones-now-hbos-most-watched-show-ever.html.

3. If the series was produced for HBO by another studio, the studio would be concerned about how the simultaneous airing would diminish later series' sales and be less likely to allow it.

4. *Game of Thrones* aired in many countries on HBO-branded channels such as HBO Asia, HBO Canada, HBO Central Europe, and HBO Latin America, but not all these channels are owned—either fully, or in part by HBO's parent company, Time Warner. Still, coordinating the simulcast with co-owned channels might not have been a considerable challenge. HBO also allowed for other channels that purchased licensing rights to the show to air it immediately, channels such as Orbit Showtime

Network, Canal+, and Orange Cinema Series—making it available in 170 countries including Africa/M-Net, Australia/Foxtel, Belgium (Flemish)/Telenet, France/Orange, Iceland/365 Media, Israel/DBS, Sky Deutschland, Sky Italia, and SKY New Zealand. HBO licensing partners in the simulcast include: Belgium (French)/Betv, Greece/Intervision, Russia/Amedia and Spain/DTS. Inexplicably the UK, serviced by Sky Atlantic, remained a day behind. [Aaron Couch, "*Game of Thrones* Season 5 Set for Global Day-Date Release," *Hollywood Reporter*, March 10, 2015; http://www.hollywoodreporter.com/live-feed/game-thrones-season-5-set-780377.]

As much as *Game of Thrones* maximized many of the possibilities of the emerging competitive environment, it offered a strategy viable in only the most exceptional of cases. HBO had new episodes of *Game of Thrones* to air only during ten weeks a year. As new subscriber-funded services such as Netflix, Hulu Plus, and HBO Now developed, the services needed competitive strategies to ensure audiences didn't subscribe merely for the three months of *Game of Thrones*. Just as internet distribution created new opportunities, it also created new challenges.

5. In fairness, the novels had been in print for more than five years, but by late seasons the narrative of the television series moved beyond that of the novels.

6. Although *Game of Thrones* seemed preordained for legendary status, its success—like all blockbusters—wasn't just a matter of the right components, but a confluence of luck and timing as well. For example, the HBO/BBC coproduction *Rome* (2005) was of similar budgetary aspiration ($100 million for twelve episodes) and ancient Rome is only slightly more culturally specific than Westeros, but *Rome* struggled and is regarded as a rare failure for HBO. [Mesce, *Inside the Rise of HBO*, 215.] *Rome* arrived too early to take advantage of digital distribution and social media, underscoring the good fortune of *Game of Thrones'* timing.

7. In truth, HBO Go did not add significant access to viewers who already had HBO On Demand, although the HBO Go interface was superior to most on-demand systems and allowed viewing on computers and mobile devices. The more significant development in this regard was HBO's 2015 launch of HBO Now, which allowed subscription to HBO independent of a multichannel provider. HBO Now circumvented the multichannel providers—who typically took half of the fee subscribers paid. What is important is that it made HBO more affordable because several cable providers required subscription to tiers of ad-supported channels that might cost $50 a month and more just to subscribe to HBO. Although available only in the United States at launch, HBO Now responded to concerns that the service was too expensive.

8. Occasionally, a well-known and sought-after writer/producer might be offered a "put pilot" as a bargaining tactic. A put pilot guaranteed a future new series that had to be scheduled.

9. Patrick Seitz, "Netflix Needs Another Hit, but *Marco Polo* Might Be All Wet," *Investor's Business Daily*, December 10, 2014.

10. Amanda D. Lotz, *Portals: A Treatise on Internet-Distributed Television* (Ann Arbor, MI: Maize Publishing, 2017); http://quod.lib.umich.edu/m/maize/mpub9699689/.

11. John Landgraf, speech, Internet and Television Expo, Chicago, IL, May 2015.

Chapter 26

1. Perhaps "independent professional" captures the commercially oriented nature of those developing profitable enterprises on YouTube without the massive infrastructure of the legacy industry.

2. YouTube has devoted funds to program development and makes available YouTube Studios for certain creators. Only a tiny percentage of the available video receives this development aid.

3. Jomar, "The Music Marketing Lesson from Beyoncé," *Music Producers Forum*, January 7, 2014; http://www.musicproducersforum.com/2014/01/07/music-marketing-lesson-from-beyonce/.

4. *The Howard Stern Show*, April 11, 2016; https://soundcloud.com/howardstern/louisck_debt

5. Hilary Lewis, "Louis C.K. Says *Horace and Pete* Coming to Hulu, Responds to Jill Soloway's Criticism of 'Trans' Scene," *Hollywood Reporter*, October 11, 2016; http://www.hollywoodreporter.com/live-feed/louis-ck-horace-pete-hulu-937633

6. Of course, Marshall Herskovitz and Edward Zwick's *Quarterlife* (2007–2008), as well as series such as *Web Therapy* (2008–2014) are significantly earlier predecessors. C.K.'s experiment differs as an hour-long drama and in his use of a personal distribution infrastructure. C.K. drew the first contours of what practices of production and release might be characteristic of this type of series. By requiring a fee to access the show he made clear that this was a commercial endeavor. C.K. significantly required a fee per episode and required a payment without sampling. In contrast to other experiments in digital distribution that allow a free trial or initial free content, the pilot fee declared that creative work has a cost.

C.K.'s experiment was also valuable and informative with regard to attitudes toward business models for internet-distributed television. It was significant that C.K. opted for a transaction payment and that he chose to launch the pilot at the above-market price of $5. C.K. had established himself as a "friend" of his fans and generous with earnings. In his earlier download sales, he emphasized that the price he sold his stand-up specials for was much lower than would be the case if a studio released them, and his release enabled downloads that would work across a range of devices and without geographic limitations. After "Live at the Beacon Theater" proved much more successful than he expected, he shared a quarter of the windfall as a bonus to those who had worked on the production and donated another quarter to five charities. C.K. consequently had cultivated a level of goodwill very different from most commercial media institutions.

Upon announcing *Horace and Pete,* C.K. addressed the price point specifically, noting that a one-hour drama was much more expensive to produce than a stand-up special. Still, the price point was not without controversy. Even though single episodes of series commonly retailed for $2.99 per episode on iTunes—which is what C.K. charged for episodes three through ten—few viewers commonly paid for television in a transaction payment, which led instead to the comparison of the per episode fee with a monthly Netflix fee of $8.

C.K.'s forthrightness about the cost of the series and his demonstration that art has a cost shifted the onus of the transaction to the audience member. Unlike so many "corporate" media products that people download illegally without much thought of consequence, not paying for *Horace and Pete* became an act against C.K. in a way important for acculturating audiences to understand that sophisticated video entertainment often carries a high creation cost. Of course *Horace and Pete* could also be downloaded illegally.

It is also important to consider the budget and its consequences. C.K.'s suggestions of profit sharing with the other actors suggests that they worked below scale and maybe only for a stake of profits (C.K.'s managers refused interview requests). This financing opportunity resulted partly from the personal relationships C.K. had cultivated as an established talent, and also from the willingness to be paid less to be involved in something of such creative aspiration as is often the case when top Hollywood talent appears in independent films. Some evidence exists that the creative opportunities of internet distribution are coming at notable financial cost for creatives, and that is a consideration that must be part of assessing these opportunities and the potential of these new distribution outlets.

The case revealed information about audience behavior. C.K. overestimated how many viewers would watch in real-time as episodes were released, especially in an environment of such programming surplus. The added experiment of eschewing prepublicity also deflated early interest and the possibility of a sizable audience watching as it was released, despite its other accomplishments. Beyond this goal held by C.K., it is difficult to know how to measure success. The metrics developed for linear distribution—such as how many paid to download episodes each week—aren't meaningful in a nonlinear environment. The measure of nonlinear, ad-free television isn't how many view in the first week, month, or even year, but how much value it accrues for the licensor or owner over its full life. The need for C.K. to be patient with revenue was no different than the deficit financing situation undergirding the studios.

The case also affirmed the challenges that are part of self-production and distribution, even for widely recognized celebrities who have the email addresses of everyone who had purchased from them. Just as in music and print, the *Horace and Pete* experiment proved that big media industries have marketing capabilities that remain valuable for helping content find audiences.

Chapter 27

1. KCRW's The Business, "Interview with John Landgraf," September 18, 2015.

2. Jeffrey A. Eisenach, "Retransmission Consent and the U.S. Market for Video Content," NERA Economic Consulting, July 2014; http://www.nera.com/publications/archive/2014/delivering-for-television-viewers-retransmission-consent-and-th.html.

3. Cecilia Kang, "The Internet Is Slowly but Surely Killing TV," *Washington Post*, September 9, 2015; https://www.washingtonpost.com/news/the-switch/wp/2015/09/09/the-internet-is-slowly-but-surely-killing-tv/.

4. Amanda D. Lotz, *Portals: A Treatise on Internet-Distributed Television* (Ann Arbor, MI: Maize Publishing, 2017); http://quod.lib.umich.edu/m/maize/mpub9699689/.

5. Analysis by Liam Boluk, "The State and Future of Netflix v. HBO in 2015," *ReDEF*, March 5, 2015; http://redef.com/original/the-state-and-future-of-netflix-v-hbo-in-2015. Netflix was arguably more diversified than HBO. In addition to its library and original streaming content, it maintained its DVD by-mail service that offered a deeper catalog and faster availability of recent films. Netflix's television offerings for kids were also often overlooked by adults assessing the service, but Netflix's creation of a walled garden of ad-free content with an interface easy enough for children to navigate was of key value to parents and practically justified the expense of the subscription on its own.

Chapter 28

1. Lacey Rose, "HBO's Real-Life *Game of Thrones* Fight to Stay Rich, on Top and Score Bill Simmons," *Hollywood Reporter*, June 17, 2015; http://www.hollywoodreporter.com/news/hbos-real-life-game-thrones-802764.

2. HBO Now fees wouldn't be shared with cable providers, but HBO initially sold the service through second parties such as Apple, Amazon, Android, and Verizon rather than developing its own customer service infrastructure. Such an arrangement warranted that providers receive some percentage of the subscription fee, although it positioned HBO to shift to a direct-to-consumer model if revenues were worthwhile.

3. "Why 8000,000 HBO Now Subscribers Isn't That Disappointing," *Motley Fool*, February 19, 2016; http://www.fool.com/investing/general/2016/02/19/why-800000-hbo-now-subscribers-isnt-that-disappoin.aspx/.

4. CBS owned only a handful of its affiliate stations; the rest were owned by others. The launch of CBS All Access thus threatened to diminish the number of viewers who watched their local CBS channel and could be sold to local advertisers. The relationship between local affiliate stations has shifted tremendously as a result of

these changes, and this too required particular negotiation—most likely CBS sharing some of the subscriber revenue.

5. Oriana Schwindt, "Here's Why CBS Is the Future of Television No One Saw Coming (Except Les Moonves)," *International Business Times*, February 12, 2016; http://www.ibtimes.com/heres-why-cbs-future-television-no-one-saw-coming-except-les-moonves-2303668; Ben Munson, "Showtime OTT Hits 1.5M Subs, CBS All Access Not Far Behind," *FierceCable*, February 14, 2017; http://www.fiercecable.com/broadcasting/showtime-ott-hits-1-5m-subs-cbs-all-access-not-far-behind

6. Jacob Kastrenakes, "Hulu Now Has 12 Million Subscribers, but Growth Is Slowing," *The Verge*, May 4, 2016; http://www.theverge.com/2016/5/4/11589886/hulu-hits-12-million-subscribers. In addition to some viewer revolt against the experience of the tediousness of ads, the ad-supported services also struggled because it was unclear how to value their commercials. Advertisers were cautious because of the considerable uncertainty about whether data measuring who watched ads and whether viewing reported by services were valid. Moreover, ads in a linear era derived their value from scarcity—there could be only so many ads, which limited supply and drove up their cost. There was no such constraint for internet distribution.

7. Hulu's niche by 2015 was as a portal for those who did not have multichannel service but wanted to remain current with Fox, NBC, and ABC shows. Its reliance on ad-support, even including ads for those who paid $8 monthly to access the most recent shows, gave it a substandard viewing experience compared with ad-free subscriber services until it launched an ad-free alternative for an additional $4 per month in summer 2015. Hulu had an incumbent's advantage when the mass market was ready for internet-distributed video in 2015 and also achieved more—but still limited—attention for its original content. In May 2016 it announced that the service would expand to offer the live feeds of various channels by internet as well.

As Hulu's shifting business model suggests, even though the arrival and expansion of "digital" advertising drew extensive attention from the inception of internet-distributed television services, by 2015 it became increasingly clear that subscriber services provided a stronger business model at this stage of the distribution technology. Other ad-supported portals, such as Sony's Crackle, had also been in the market since 2007, but the available ad-dollars couldn't offset the cost of programming that would attract audiences. Ad-supported portals consequently failed to garner as much interest from viewers as subscriber-supported Netflix. Online video ads still only accounted for a tenth of the $70 billion television advertising market in 2015. [Steve Lohr, "With the TV Business in Upheaval, Targeted Ads Offer Hope," *New York Times*, September 29, 2015.]

8. Shalini Ramachandran, "Want to Watch Zombies 24/7? There's a Pay Site for That," *Wall Street Journal*, April 12, 2016.

9. Sarah Perez, "Disney to Launch a Subscription Streaming Service in Europe," *Tech Crunch*, October 21, 2015; http://techcrunch.com/2015/10/21/disney-to-launch-a-subscription-streaming-service-in-europe-hints-that-marvel-star-wars-services-could-follow-in-u-s/.

10. Aaron Baar, "Connected TV Devices Reach Tipping Point," *Media Post*, August 31, 2015; http://www.mediapost.com/publications/article/257343/connected-tv-devices-reach-tipping-point.html. The cable industry's TV Everywhere initiative also enabled multidevice access, but the internet portals enabled consumers to begin constructing an à la carte set of options that was still nearly impossible through cable subscription. By the end of 2015, it was possible to legally access a considerable amount of current television—except sports events—through an internet-distributed service, especially if viewers were willing to delay viewing.

Chapter 29

1. At launch: ESPN, ESPN2, TNT, TBS, Food Network, HGTV, Travel Channel, Adult Swim, Cartoon Network, Disney Channel, ABC Family, and CNN. Most notably, Sling TV's basic package included ESPN. Access to sports kept many homes reliant on cable because most events were available only from cable channels and could not be streamed. This was the first fracture in the requirement of multichannel service to access ESPN and its major sports licenses.

2. In March 2015, Sony dropped the price $10 per month to $40 for sixty channels and added ABC channels including ESPN in the cheapest tier. [Mike Snider, "PlayStation Vue Gets Price Cut Plus ABC, ESPN," *USA Today*, March 2, 2016; http://www.usatoday.com/story/tech/news/2016/03/02/playstation-vue-gets-price-cut-plus-abc-espn/81201964.]

3. "Number of Sling TV Subscribers as of September 2016," Statista, https://www.statista.com/statistics/539242/sling-tv-subscribers/; Jason Dunning, "Report: PlayStation Vue Has Over 100,000 Subscribers," *PlayStation LifeStyle*, July 4, 2016; http://www.playstationlifestyle.net/2016/07/04/report-playstation-vue-100000-subscribers/.

4. Daniel Frankel, "Sling TV Now Has More Than 2M Subscribers, comScore Says," *Fierce Cable*, June 23, 2017; http://www.fiercecable.com/cable/sling-tv-has-more-than-2m-subs-comscore-says

4. Five broadcast networks, CNN, HGTV, AMC, Food Network, WE, Univision, and Telemundo and twenty-three obscure channels. The Verizon bundle didn't seem a game changer because the service was available only to homes wired for FiOS, but it initially included ESPN in one of the packs and provided a more precise option for viewers with specific tastes. The service drew significant attention when ESPN sued, arguing that Verizon violated its contract by placing it in a "sports pack," but

it added momentum to the repackaging of television services that occurred throughout 2015. Nonetheless, a revised version of the service launched in February 2016 that offered a Custom TV-Essentials package, including broadcast networks and seventy-eight cable channels for $55 a month and a Custom TV- Sports & More package for $70 for those seeking ESPN; these packages quickly escalated in price. [Todd Spangler, "Verizon Revamps 'Skinny' FiOS TV Bundles After Spats with ESPN, Others," *Variety*, February 19, 2016; http://variety.com/2016/tv/news/verizon-fios-custom-tv-revamp-espn-lawsuit-1201710497/.]

6. Notably, with the exception of Sling TV's nationwide availability, most of the "choice" was isolated to the major cities that already enjoyed competition from two wired providers (traditional cable and telco). The competitiveness of major urban markets had grown, but nonmetropolitan-area America remained under monopoly control and was the last place new innovations were implemented. Cities also had more abundant internet infrastructure necessary for streaming large amounts of video. Despite all the things internet connections enabled as the backbone of twenty-first century modernity, it was the leisure behavior of television viewing that provided the most urgent motivation for innovation. During the first decade of the twenty-first century, demand for internet-delivered video grew despite very limited content availability. Once business models for packaging content emerged in 2015, a surplus of programming became accessible. The question then became whether the infrastructure existed for a majority of viewers to migrate to new viewing behaviors.

Chapter 30

1. Daniel Frankel, "Cable's Ratings Collapse," *Fierce Cable*, March 12, 2015.

2. Cynthia Littleton, "TV's New Math: Networks Crunch Their Own Ratings to Track Multiplatform Viewing," *Variety*, February 11, 2015; http://variety.com/2015/tv/features/broadcast-nets-move-closer-to-developing-ratings-that-consider-auds-delayed-viewing-habits-1201430321/.

3. Nielsen, prepanel slides at NATPE 2015.

4. Littleton, "TV's New Math."

5. Mike Farrell, "Making Multiplatform Viewing Count," *Multichannel News,* April 13, 2015, 4–5, 4.

6. Based on Netflix subscribers spending an average of 1.5 hours a day streaming from the service, and an average hour of television having approximately 15 minutes and 30 seconds of ads. [Rob Toledo, "Netflix Saves Us from Over 130 Hours of Commercials a Year," *Exstreamist,* August 17, 2015; http://exstreamist.com/netflix-saves-us-from-over-130-hours-of-commercials-a-year/].

7. Alan Wurtzel, quoted in Farrell, "Making Multiplatform Viewing Count."

8. John Lafayette, "Netflix Main Cause of TV Ratings Drop," *Broadcasting & Cable*, April 23, 2015; http://www.broadcastingcable.com/news/currency/netflix-main-cause-tv-ratings-drop/140194.

9. On-demand viewing done within three days of airing counted in the ratings figure, but the majority of viewing took place after day three, and more than 50 percent occurred at least seven days after airing. The panic over the disappearing live audience led networks and channels to push for changes in the measurements on which advertising sales are based. By the time the channels and networks began negotiating advertising commitments for 2015–2016, many were pushing to count viewing within seven days, and some even advocated expanding the window to thirty days. The news wasn't uniformly terrible if audiences were counted differently. Viewers were still watching, just not in ways that supported broadcast era norms. [Rentrak, "Video on Demand Broadcast Primetime Viewing Grew 22 Percent According to the Rentrak 'State of VOD: Trend Report,'" *Rentrak Press Release*, June 15, 2015; http://www.prnewswire.com/news-releases/video-on-demand-broadcast-primetime-viewing-grew-22-percent-according-to-the-rentrak-state-of-vod-trend-report-300098780.html.] Notably, Comcast data released in the wake of a massive 2015 snowstorm that a record of 1.29 million concurrent streams was set during the storm suggests less VOD use at any given moment than might be assumed—1.29 million amounts to just 5 percent of Comcast's subscribers. [Jeff Baumgartner, "Comcast: VOD Usage Surges During Jonas," *Multichannel News*, January 25, 2016; http://www.multichannel.com/news/vod/comcast-vod-usage-surges-during-jonas/396818.]

A key part of the problem was that the legacy industry tried to impose the practices of the broadcast/cable paradigm on internet distribution even when it attempted experiments with internet distribution, for example, the strategies used with Hulu. Audiences were still being sold as giant demographic segments; commercial pods filled 30 percent of a program hour; audiences for shows were based on their viewing at specific times. None of these features incorporated the different practices made possible by internet distribution. Despite this obvious mismatch of legacy advertising practices and a new mechanism of distribution, neither the channels nor the advertisers endeavored to lead the way to new mutually beneficial norms.

Conclusion

1. The publishing industry maintains a windowing practice of publishing books expected to have high sales first in hardback. The changes in the retailing practices of books have largely eroded the premium rate that made driving readers to this window worthwhile.

2. The value of internet-distributed television services derives from what content they provide, the quality of the experience, and the value proposition. Top services must allow flexibility across devices, and provide interfaces, search functionality, and recommendations that assist viewers' navigation of the surplus of content. Services that offer a range of price points—an advertising subsidized and a subscriber-supported option—better allow viewers to find a suitable value proposition. With a surplus of content available, viewers' selection of what to watch combines an assessment of the content and the viewing experience. Those that attempt to enforce certain viewing conditions—such as watching ads—risk losing viewers interested in their content to preferable viewing conditions. Likewise, surplus prevents viewers from being held hostage to exclusivity. Content made difficult to access owing to price point or viewing condition goes unwatched in this environment. The impossibility of perfect substitution makes exclusivity a viable strategy in few cases—only the NFL portal will have the NFL and nothing else is a true substitute; or only HBO has *Game of Thrones* and no other show can be a perfect substitute—but only the most prized content will warrant unreasonable rates for access, and little content is likely to be widely prized.

If the legacy television industry follows the strategy of studio portals started by HBO and CBS, it has the capability to accomplish something yet to be achieved among media migrating to internet distribution. Every legacy media entity has missed the opportunity of vertically integrating production and distribution to develop relationships directly with consumers. The music labels didn't develop their own platform for audiences to listen to the music of their artists, they waited for an outsider—Apple—to come in, become the dominant player, and take a big cut of revenue. The film industry, likewise, has allowed distributors such as Netflix, iTunes, and cable providers to most effectively distribute its content after theatrical release. Newspaper publishing has come closest to self-distribution, but it has struggled with the business model. Launching studio portals is a costly risk, but it also provides the potential for considerable reward and full supply chain control.

3. See Stuart Cunningham, David Craig, and Jon Silver, "YouTube, Multichannel Networks and the Accelerated Evolution of the New Screen Ecology," *Convergence* 22, no. 4 (2016): 376–391 and emerging work from David Craig and Stuart Cunningham.

4. It is not clear which of the emerging internet-distributed competitors or practices will endure, although evidence of many options abounds. Ad-supported, video on demand provides the most viable option and entails limited change, but its success will require fast improvement of dynamic ad insertion and adjustment in advertising metrics and practices. These audiences will remain only if costs and pricing logics for cable packages respond and allow audiences to achieve a better value proposition.

The problem for advertising is that the audiences many advertisers seek—the young or more affluent—will likely continue their migration away from ad-supported television. A lot remains up for grabs: will more internet-distributed subscription services emerge; will ad-supported television reduce its commercial load; will technologies of internet distribution continue to advance at a rate fast enough to accommodate a majority of viewers choosing to access content this way; will new administrations overturn net neutrality policy; will the monopolists of cable, and now internet, service maintain their monopolies and reputations of poor service; will they force pricing of internet that discourages extensive internet streaming; when will cellular and satellite develop adequate capability to distribute high quality, long-form video content and business models that allow prevalent use?

Index